KUMON MATH WORKBOOKS

Pre-Algebra

Workbook I

Table of Contents

KUMON

1 Fraction Review

Date / /

Name

Level ☆

Score /100

■ The Answer Key is on page 88.

1 Rewrite each improper fraction as a mixed number or whole number. 2 points per question

(1) $\frac{5}{4} = 1\frac{1}{4}$

(2) $\frac{7}{5} =$

(3) $\frac{4}{4} =$

(4) $\frac{7}{3} =$

(5) $\frac{7}{4} =$

(6) $\frac{6}{3} =$

(7) $\frac{5}{2} =$

(8) $\frac{9}{4} =$

(9) $\frac{10}{3} =$

(10) $\frac{15}{11} =$

(11) $\frac{9}{9} =$

(12) $\frac{13}{7} =$

(13) $\frac{14}{7} =$

(14) $\frac{12}{3} =$

(15) $\frac{11}{4} =$

(16) $\frac{13}{3} =$

(17) $\frac{15}{7} =$

(18) $\frac{3}{1} =$

(19) $\frac{17}{9} =$

(20) $\frac{18}{7} =$

2 Rewrite each whole number as a fraction.

3 points per question

(1) $1=\frac{4}{4}$

(2) $1=\frac{\square}{7}$

(3) $2=\frac{\square}{5}$

(4) $2=\frac{\square}{7}$

(5) $3=\frac{\square}{3}$

(6) $3=\frac{\square}{4}$

(7) $1=\frac{\square}{9}$

(8) $2=\frac{\square}{8}$

3 Rewrite each mixed number as an improper fraction.

3 points per question

(1) $1\frac{1}{2}=\frac{3}{2}$

(2) $1\frac{4}{5}=$

(3) $2\frac{4}{5}=$

(4) $2\frac{5}{6}=$

(5) $1\frac{4}{7}=$

(6) $2\frac{1}{3}=$

(7) $4\frac{2}{3}=$

(8) $1\frac{7}{13}=$

(9) $3\frac{5}{11}=$

(10) $2\frac{1}{8}=$

(11) $1\frac{4}{11}=$

(12) $5\frac{3}{4}=$

A fraction such as $\frac{3}{2}$, whose numerator is greater than the denominator is called an improper fraction.

Great start!

2 Reduction Review

Date / /

Name

■ The Answer Key is on page 88.

1 Reduce by dividing the numerator and denominator by 2, 3, or 5.

2 points per question

(1) $\frac{3}{6} = \frac{1}{2}$

(2) $\frac{5}{15} =$

(3) $\frac{12}{15} =$

(4) $\frac{14}{20} =$

(5) $\frac{21}{30} =$

2 Reduce by dividing the numerator and denominator by 2, 3, or 7.

2 points per question

(1) $\frac{7}{21} =$

(2) $\frac{12}{14} =$

(3) $\frac{21}{35} =$

(4) $\frac{12}{21} =$

(5) $\frac{30}{33} =$

Reduction means simplifying a fraction by dividing the numerator and denominator by the same number.

3 Reduce.

4 points per question

Example $\frac{16}{28}=\frac{8}{14}=\frac{4}{7}$

(1) $\frac{8}{16}=$

(2) $\frac{12}{16}=$

(3) $\frac{25}{35}=$

(4) $\frac{15}{21}=$

(5) $\frac{30}{36}=$

(6) $\frac{10}{30}=$

(7) $\frac{20}{44}=$

(8) $\frac{14}{56}=$

You may want to reduce more than once.

4 Reduce.

4 points per question

(1) $\frac{12}{20}=$

(2) $\frac{20}{24}=$

(3) $\frac{8}{56}=$

(4) $\frac{14}{28}=$

(5) $\frac{12}{36}=$

(6) $\frac{25}{50}=$

(7) $\frac{9}{21}=$

(8) $\frac{28}{60}=$

(9) $\frac{25}{75}=$

(10) $\frac{22}{66}=$

(11) $\frac{18}{45}=$

(12) $\frac{21}{56}=$

Try to reduce each fraction in one step.

If you have to reduce in more than one step, just keep practicing. You'll get it!

3 Greatest Common Factor

Level ☆

Date / /

Name

■ The Answer Key is on page 88.

Don't forget!

8 can be divided evenly by 1, 2, 4, and 8. This means that 1, 2, 4, and 8 are **factors** of 8.

1 Write the appropriate number in each box.

4 points per question

(1) The factors of 16 are … 1, 2, 4, [8], 16

(2) The factors of 20 are … 1, 2, 4, [ ], 10, 20

(3) The common factors of 16 and 20 are … 1, 2, [ ]

(4) The factors of 12 are … 1, 2, 3, 4, [ ], 12

(5) The factors of 24 are … 1, 2, 3, 4, 6, 8, [ ], 24

(6) The common factors of 12 and 24 are … 1, 2, 3, 4, [ ], 12

Don't forget!

The factors that two or more integers have in common are called **common factors**.

2 Write the appropriate number in each box.

2 points per question

(1) The greatest common factor of 16 and 20 is [4].

(2) The greatest common factor of 12 and 24 is [ ].

Don't forget!

Among the common factors, the largest factor that two integers have in common is called the **greatest common factor** (**GCF**).

3 Write the GCF of each number pair.

3 points per question

Example (16, 20) → [4]

(1) (12, 30) → [ ]

(2) (20, 45) → [ ]

(3) (14, 42) → [ ]

(4) (45, 54) → [ ]

4 **Find the GCF of each number pair. Then reduce each fraction by the GCF.**

3 points per question

Example (18, 27) → GCF [9] $\frac{18}{27}=\frac{2}{3}$

(1) (12, 24) → [ ] $\frac{12}{24}=$

(2) (16, 20) → [ ] $\frac{16}{20}=$

(3) (16, 28) → [ ] $\frac{16}{28}=$

(4) (14, 35) → [ ] $\frac{14}{35}=$

(5) (15, 30) → [ ] $\frac{15}{30}=$

(6) (10, 40) → [ ] $\frac{10}{40}=$

(7) (18, 36) → [ ] $\frac{18}{36}=$

(8) (9, 33) → [ ] $\frac{9}{33}=$

(9) (8, 56) → [ ] $\frac{8}{56}=$

(10) (9, 54) → [ ] $\frac{9}{54}=$

(11) (36, 48) → [ ] $\frac{36}{48}=$

(12) (20, 70) → [ ] $\frac{20}{70}=$

(13) (22, 55) → [ ] $\frac{22}{55}=$

(14) (35, 60) → [ ] $\frac{35}{60}=$

(15) (23, 46) → [ ] $\frac{23}{46}=$

(16) (25, 65) → [ ] $\frac{25}{65}=$

(17) (14, 21) → [ ] $\frac{14}{21}=$

(18) (18, 45) → [ ] $\frac{18}{45}=$

(19) (20, 90) → [ ] $\frac{20}{90}=$

(20) (29, 87) → [ ] $\frac{29}{87}=$

You are the greatest at finding common factors!

4 Comparing Fractions

Level  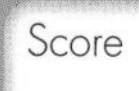

Date / /

Name

Score /100

■ The Answer Key is on page 88.

Don't forget!

The symbol > is read as "is greater than," and the symbol < is read as "is less than."

1 **Reduce each pair of fractions and compare.**

4 points per question

(1) $\frac{12}{15}=\frac{4}{5}$, $\frac{6}{10}=\frac{3}{5}$

$\frac{4}{5} > \frac{3}{5}$

(2) $\frac{7}{70}=\frac{\square}{\square}$, $\frac{9}{30}=\frac{\square}{\square}$

$\text{—} \square \text{—}$

(3) $\frac{55}{75}=\text{—}$, $\frac{14}{30}=\text{—}$

$\text{—} \square \text{—}$

(4) $\frac{15}{27}=\text{—}$, $\frac{48}{54}=\text{—}$

$\text{—} \square \text{—}$

(5) $\frac{24}{56}=\text{—}$, $\frac{6}{14}=\text{—}$

$\text{—} \square \text{—}$

(6) $\frac{8}{52}=\text{—}$, $\frac{32}{52}=\text{—}$

$\text{—} \square \text{—}$

(7) $\frac{3}{18}=\text{—}$, $\frac{25}{30}=\text{—}$

$\text{—} \square \text{—}$

(8) $\frac{20}{35}=\text{—}$, $\frac{44}{77}=\text{—}$

$\text{—} \square \text{—}$

(9) $\frac{15}{27}=\text{—}$, $\frac{14}{63}=\text{—}$

$\text{—} \square \text{—}$

(10) $\frac{28}{48}=\text{—}$, $\frac{22}{24}=\text{—}$

$\text{—} \square \text{—}$

2 Compare each pair of fractions.

5 points per question

(1) $\frac{20}{30}$, $\frac{20}{60}$

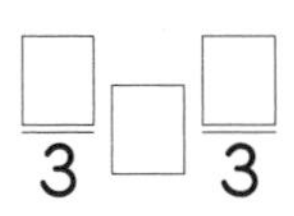

(2) $\frac{18}{45}$, $\frac{10}{25}$

—□—

(3) $\frac{8}{32}$, $\frac{12}{16}$

—□—

(4) $\frac{24}{52}$, $\frac{6}{39}$

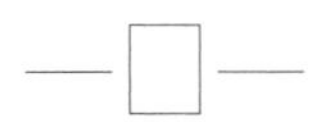

(5) $\frac{50}{80}$, $\frac{24}{64}$

—□—

(6) $\frac{32}{56}$, $\frac{12}{21}$

—□—

To compare two fractions, first make their denominators equal.

3 Answer each word problem.

15 points per question

(1) Amanda has $\frac{15}{16}$ pound of beads. Tessa has $\frac{24}{32}$ pound of beads. Who has more beads?

$\frac{15}{16}$, $\frac{24}{32}$ → —□—

⟨**Ans.**⟩

(2) Leslie ate $\frac{12}{36}$ pint of blueberries. Joey ate $\frac{18}{27}$ pint of strawberries. Who ate more fruit?

⟨**Ans.**⟩

5 Least Common Multiple

Level ☆

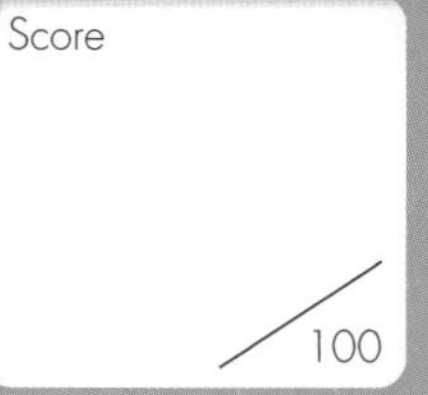
Score /100

Date / / Name

■The Answer Key is on page 88.

1 **Write the multiples of 4 in ascending order.** 4 points for completion

4, 8, 12, [16], [ ], [ ], [ ], [ ], [ ], …

2 **Write the multiples of 6 in ascending order.** 4 points for completion

6, 12, 18, [ ], [ ], [ ], [ ], [ ], [ ], …

3 **Write the multiples of 7 in ascending order.** 4 points for completion

7, 14, [ ], [ ], [ ], [ ], [ ], [ ], [ ], …

4 **Write the multiples of 9 in ascending order.** 4 points for completion

9, [ ], [ ], [ ], [ ], [ ], [ ], [ ], [ ], …

5 **Write the common multiples of 4 and 6 in ascending order.** 4 points for completion

[12], [ ], [ ], …

Don't forget!
The multiples that two integers have in common are called **common multiples**.

6 **Write the common multiples of 6 and 9 in ascending order.** 4 points for completion

[ ], [ ], [ ], …

Don't forget!
The smallest common multiple is called the **least common multiple (LCM)**.

7 **Write the common multiples of 4 and 7 in ascending order.** 4 points for completion

[ ], …

8 Write the appropriate number in each box.

2 points per question

(1) The least common multiple of 4 and 6 is ... ☐

(2) The least common multiple of 6 and 9 is ... ☐

(3) The least common multiple of 4 and 7 is ... ☐

(4) The least common multiple of 4 and 9 is ... ☐

9 Find the LCM of each number pair.

4 points per question

Example (4, 6) → 12

(1) (6, 8) → ☐

(2) (4, 10) → ☐

(3) (2, 9) → ☐

(4) (6, 12) → ☐

(5) (3, 8) → ☐

(6) (3, 4) → ☐

(7) (8, 12) → ☐

(8) (5, 7) → ☐

(9) (9, 12) → ☐

(10) (6, 10) → ☐

(11) (7, 8) → ☐

(12) (9, 15) → ☐

(13) (3, 11) → ☐

(14) (2, 7) → ☐

(15) (7, 9) → ☐

(16) (5, 12) → ☐

Remarkable job!

6 Comparing Fractions

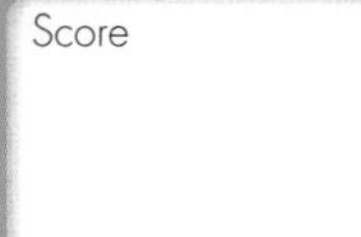

Date / /　Name

■ The Answer Key is on page 89.

1 Find the LCM of the denominators in each pair of fractions. Then compare the fractions.

5 points per question

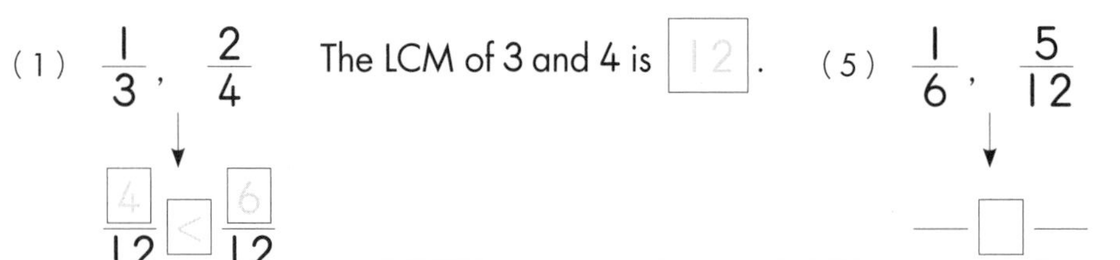

(1) $\frac{1}{3}$, $\frac{2}{4}$　The LCM of 3 and 4 is 12.

$\frac{4}{12} < \frac{6}{12}$

(5) $\frac{1}{6}$, $\frac{5}{12}$　The LCM of the denominators is ☐.

— ☐ —

(2) $\frac{3}{4}$, $\frac{7}{10}$　The LCM of the denominators is 20.

— ☐ —

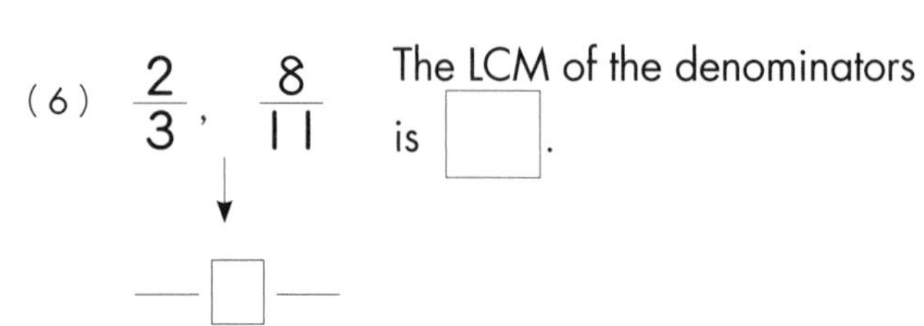

(6) $\frac{2}{3}$, $\frac{8}{11}$　The LCM of the denominators is ☐.

— ☐ —

(3) $\frac{3}{8}$, $\frac{5}{12}$　The LCM of the denominators is ☐.

— ☐ —

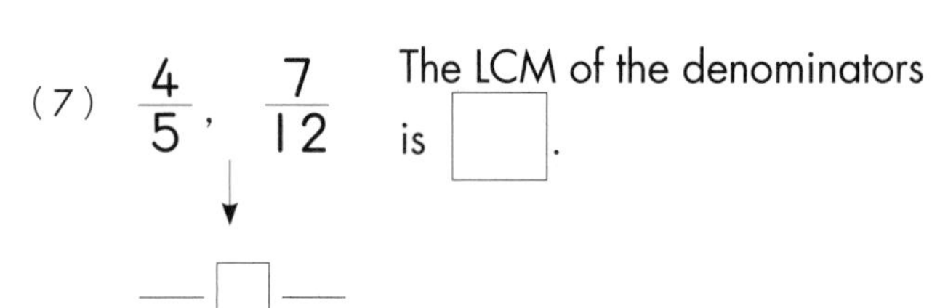

(7) $\frac{4}{5}$, $\frac{7}{12}$　The LCM of the denominators is ☐.

— ☐ —

(4) $\frac{2}{5}$, $\frac{6}{7}$　The LCM of the denominators is ☐.

— ☐ —

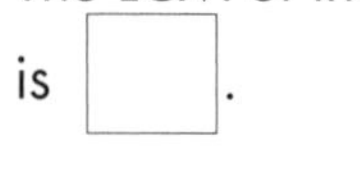

(8) $\frac{4}{5}$, $\frac{9}{10}$　The LCM of the denominators is ☐.

— ☐ —

To compare two fractions, first make their denominators equal.

2 Compare each pair of fractions.

5 points per question

(1) $\frac{1}{4}$, $\frac{5}{6}$

(2) $\frac{3}{5}$, $\frac{7}{9}$

— □ —

(3) $\frac{8}{9}$, $\frac{12}{15}$

— □ —

(4) $\frac{5}{6}$, $\frac{7}{9}$

— □ —

(5) $\frac{3}{4}$, $\frac{3}{7}$

— □ —

(6) $\frac{1}{2}$, $\frac{3}{13}$

— □ —

3 Answer each word problem.

15 points per question

(1) Sarah is $\frac{5}{7}$ as tall as her mother. Her sister Leah is $\frac{10}{11}$ as tall as their mother. Who is taller, Leah or Sarah?

〈**Ans.**〉

(2) Mr. Hall's gym class is $\frac{3}{4}$ full. Mrs. Castamore's dance class is $\frac{8}{9}$ full. Who's class is more full?

〈**Ans.**〉

You can compare any fractions! Well done!

7 Addition of Fractions

Level ☆

Date / /

Name

Score /100

■ The Answer Key is on page 89.

1 Add.

3 points per question

Examples $\frac{3}{8}+\frac{3}{8}=\frac{6}{8}=\frac{3}{4}$ $\frac{3}{8}+\frac{5}{8}=\frac{8}{8}=1$ $\frac{7}{8}+\frac{3}{8}=\frac{10}{8}=1\frac{2}{8}=1\frac{1}{4}$

(1) $\frac{3}{5}+\frac{1}{5}=$

(2) $\frac{5}{8}+\frac{3}{8}=\frac{8}{8}=$

(3) $\frac{5}{7}+\frac{4}{7}=\frac{9}{7}=$

(4) $\frac{5}{9}+\frac{8}{9}=$

(5) $\frac{5}{6}+\frac{5}{6}=\frac{10}{6}=1\frac{4}{6}=$

(6) $\frac{7}{12}+\frac{11}{12}=$

(7) $\frac{7}{9}+\frac{1}{9}=$

(8) $\frac{6}{11}+\frac{7}{11}=$

(9) $\frac{9}{13}+\frac{4}{13}=$

(10) $\frac{11}{12}+\frac{5}{12}=$

(11) $\frac{9}{10}+\frac{9}{10}=$

(12) $\frac{7}{8}+\frac{5}{8}=$

Always reduce and write your answers as proper fractions.

2 Find the LCM of the denominators in each pair of fractions.

4 points per question

(1) $\frac{1}{6}, \frac{8}{9}$ → 18

(2) $\frac{2}{3}, \frac{2}{5}$ →

(3) $\frac{1}{4}, \frac{5}{9}$ →

(4) $\frac{5}{6}, \frac{2}{15}$ →

(5) $\frac{3}{4}, \frac{13}{16}$ →

(6) $\frac{4}{5}, \frac{5}{6}$ →

(7) $\frac{3}{8}, \frac{7}{10}$ →

(8) $\frac{1}{6}, \frac{13}{14}$ →

(9) $\frac{3}{10}, \frac{12}{15}$ →

(10) $\frac{8}{9}, \frac{5}{12}$ →

3 Find the LCM and then use it to add the fractions.

4 points per question

(1) (3, 4) → 12

$\frac{1}{3}+\frac{1}{4}=\frac{4}{12}+\frac{3}{12}=$

(2) (8, 12) →

$\frac{3}{8}+\frac{5}{12}=$

(3) (7, 8) →

$\frac{3}{7}+\frac{3}{8}=$

(4) (4, 9) →

$\frac{1}{4}+\frac{5}{9}=$

(5) (6, 8) →

$\frac{5}{6}+\frac{1}{8}=$

(6) (9, 15) →

$\frac{7}{9}+\frac{2}{15}=$

This is tough work, but you are doing great.

Addition of Fractions

Date / /

Name

■ The Answer Key is on page 89.

1 **Find the LCM of the denominators of the fractions and then add.** 4 points per question

Example $\frac{1}{3}+\frac{1}{7}=\frac{7}{21}+\frac{3}{21}=\frac{10}{21}$

(1) $\frac{2}{3}+\frac{1}{7}=$

(2) $\frac{3}{5}+\frac{1}{6}=$

(3) $\frac{5}{8}+\frac{3}{7}=$

(4) $\frac{5}{7}+\frac{5}{6}=$

(5) $\frac{5}{7}+\frac{4}{5}=$

(6) $\frac{7}{12}+\frac{1}{3}=$

(7) $\frac{4}{9}+\frac{1}{6}=$

(8) $\frac{4}{5}+\frac{3}{8}=$

(9) $\frac{1}{6}+\frac{5}{12}=$

(10) $\frac{11}{12}+\frac{5}{8}=$

(11) $\frac{3}{4}+\frac{2}{7}=$

(12) $\frac{3}{5}+\frac{7}{10}=$

2 Add.

4 points per question

(1) $\frac{4}{5}+\frac{1}{3}=$

(2) $\frac{2}{15}+\frac{7}{10}=$

(3) $\frac{2}{3}+\frac{6}{7}=$

(4) $\frac{5}{8}+\frac{1}{2}=$

(5) $\frac{11}{12}+\frac{1}{8}=$

(6) $\frac{3}{8}+\frac{5}{12}=$

(7) $\frac{3}{4}+\frac{2}{3}=$

(8) $\frac{1}{5}+\frac{1}{4}=$

(9) $\frac{3}{5}+\frac{9}{10}=$

(10) $\frac{3}{8}+\frac{1}{7}=$

(11) $\frac{4}{7}+\frac{1}{6}=$

(12) $\frac{1}{3}+\frac{8}{15}=$

(13) $\frac{1}{8}+\frac{3}{4}=$

Find the LCM of the denominators before adding.

Aim for perfection!

9 Addition of Fractions

Level ★★

Date / / Name

Score /100

■ The Answer Key is on page 89.

1 Add.

5 points per question

(1) $\frac{1}{6}+\frac{1}{3}+\frac{1}{2}=\frac{\square}{12}+\frac{\square}{12}+\frac{\square}{12}$

$=\frac{\square}{12}=$

(2) $\frac{1}{2}+\frac{3}{5}+\frac{1}{3}=$

(3) $\frac{4}{9}+\frac{1}{3}+\frac{5}{6}=$

(4) $\frac{4}{5}+\frac{2}{3}+\frac{11}{15}=$

(5) $\frac{4}{9}+\frac{2}{3}+\frac{5}{6}=$

(6) $\frac{2}{3}+\frac{1}{5}+\frac{1}{4}=$

(7) $\frac{3}{4}+\frac{1}{9}+\frac{1}{6}=$

(8) $\frac{3}{8}+\frac{4}{5}+\frac{1}{4}=$

(9) $\frac{3}{11}+\frac{3}{4}+\frac{1}{2}=$

(10) $\frac{1}{2}+\frac{2}{5}+\frac{3}{10}=$

Find the LCM of the denominators before adding two or more fractions with different denominators.

2 Add.

5 points per question

(1) $\frac{3}{5}+\frac{1}{2}+\frac{5}{6}=$

(2) $1\frac{1}{2}+3\frac{3}{4}+\frac{1}{8}=1\frac{\square}{8}+3\frac{\square}{8}+\frac{1}{8}$

$=$

(3) $1\frac{5}{12}+\frac{3}{4}+1\frac{1}{2}=$

(4) $1\frac{3}{10}+2\frac{1}{7}+\frac{2}{5}=$

(5) $1\frac{4}{5}+3\frac{3}{8}+4\frac{1}{4}=$

(6) $1\frac{1}{3}+2\frac{3}{4}+\frac{3}{8}=$

(7) $2\frac{1}{2}+\frac{3}{4}+1\frac{5}{6}=$

(8) $\frac{2}{3}+1\frac{1}{4}+2\frac{1}{6}=$

(9) $1\frac{3}{4}+\frac{1}{6}+2\frac{2}{9}=$

(10) $\frac{5}{6}+1\frac{3}{8}+1\frac{5}{12}=$

If you are having difficulty adding fractions, try Kumon's *Grade 6 Fractions* workbook for extra practice.

You are great at adding fractions!

10 Subtraction of Fractions

Level ★★

Date / /

Name

Score /100

■ The Answer Key is on page 89.

1 Subtract.

5 points per question

(1) $\frac{3}{7}-\frac{1}{7}=\frac{2}{7}$

(2) $\frac{3}{4}-\frac{1}{4}=$

(3) $\frac{5}{6}-\frac{1}{6}=$

(4) $\frac{8}{13}-\frac{3}{13}=$

(5) $\frac{9}{11}-\frac{4}{11}=$

(6) $\frac{4}{5}-\frac{3}{5}=$

(7) $\frac{7}{12}-\frac{5}{12}=$

(8) $\frac{11}{15}-\frac{6}{15}=$

(9) $\frac{17}{20}-\frac{11}{20}=$

(10) $\frac{23}{24}-\frac{13}{24}=$

Don't forget to reduce your answer whenever possible.

2 Subtract.

5 points per question

(1) $\frac{2}{3}-\frac{1}{4}=\frac{8}{12}-\frac{3}{12}=$

(2) $\frac{7}{12}-\frac{1}{3}=$

(3) $\frac{7}{11}-\frac{1}{2}=$

(4) $2\frac{7}{11}-1\frac{1}{2}=2\frac{\square}{22}-1\frac{\square}{22}=$

(5) $3\frac{5}{7}-1\frac{1}{3}=$

(6) $4\frac{7}{12}-2\frac{1}{2}=$

(7) $2\frac{3}{5}-1\frac{3}{10}=$

(8) $3\frac{3}{4}-1\frac{5}{12}=$

(9) $2\frac{4}{9}-\frac{1}{6}=$

(10) $3\frac{5}{6}-2\frac{3}{10}=$

Find the LCM of the denominators before subtracting two or more fractions with different denominators.

You make subtracting fractions look easy!

Subtraction of Fractions

Level ★★

Date / /

Name

Score /100

■ The Answer Key is on page 90.

1 Subtract.

5 points per question

Don't forget!

You can borrow from the whole number if the numerator of a mixed number is not large enough.

Example $3\frac{3}{4}-\frac{4}{5}=3\frac{15}{20}-\frac{16}{20}=2\frac{35}{20}-\frac{16}{20}=2\frac{19}{20}$

(1) $3\frac{1}{4}-1\frac{3}{5}=3\frac{\square}{20}-1\frac{\square}{20}$

$=2\frac{\square}{20}-1\frac{\square}{20}=$

(2) $5\frac{1}{3}-2\frac{3}{7}=$

(3) $6\frac{2}{7}-2\frac{1}{2}=$

(4) $7\frac{1}{2}-4\frac{4}{5}=$

(5) $8\frac{1}{11}-5\frac{1}{2}=$

(6) $3\frac{5}{13}-1\frac{1}{2}=$

(7) $9\frac{1}{2}-4\frac{11}{16}=$

(8) $4\frac{4}{11}-3\frac{3}{4}=$

(9) $2\frac{1}{7}-\frac{11}{14}=$

(10) $1\frac{1}{4}-\frac{9}{10}=$

2 Subtract.

5 points per question

(1) $\frac{11}{12}-\frac{1}{4}-\frac{1}{3}=\frac{\square}{12}-\frac{\square}{12}-\frac{\square}{12}$

$=$

(2) $\frac{13}{14}-\frac{1}{2}-\frac{2}{7}=$

(3) $3\frac{7}{8}-1\frac{1}{4}-\frac{1}{3}=$

(4) $7\frac{1}{8}-2\frac{2}{3}-1\frac{1}{6}=$

(5) $10\frac{7}{10}-4\frac{1}{5}-\frac{3}{4}=$

(6) $9\frac{2}{9}-3\frac{1}{2}-3\frac{1}{3}=$

(7) $5\frac{7}{8}-2\frac{1}{2}-1\frac{1}{3}=$

(8) $4\frac{3}{4}-1\frac{1}{6}-\frac{3}{8}=$

(9) $5\frac{3}{4}-1\frac{4}{9}-2\frac{1}{6}=$

(10) $3\frac{3}{8}-\frac{5}{6}-1\frac{2}{3}=$

Always calculate from left to right.

I'm impressed!

12 Addition & Subtraction of Fractions

Level 

Date / /

Name

Score /100

■ The Answer Key is on page 90.

1 **Calculate.**

5 points per question

(1) $\frac{4}{9}+\frac{1}{2}-\frac{2}{3}=\frac{\square}{18}+\frac{\square}{18}-\frac{\square}{18}$

$=$

(2) $\frac{7}{12}-\frac{1}{4}+\frac{1}{3}=$

(3) $3\frac{8}{9}-\frac{2}{3}+\frac{5}{6}=$

(4) $\frac{5}{12}+2\frac{5}{6}-\frac{3}{4}=$

(5) $5\frac{1}{9}-4\frac{1}{3}+8\frac{2}{5}=$

(6) $6\frac{1}{6}-\frac{9}{10}+\frac{1}{3}=$

(7) $15\frac{11}{12}+\frac{1}{6}-8\frac{3}{4}=$

(8) $2\frac{1}{2}-\frac{2}{3}-\frac{1}{6}=$

(9) $12\frac{1}{9}-\frac{1}{3}-\frac{1}{2}=$

(10) $4\frac{9}{14}-2\frac{6}{7}+8\frac{5}{6}=$

2 Calculate.

5 points per question

(1) $\frac{5}{12}-\frac{3}{8}+\frac{5}{6}=$

(2) $4-1\frac{2}{3}+\frac{1}{4}=$

(3) $5\frac{1}{2}-\frac{5}{6}+2\frac{2}{3}=$

(4) $6\frac{1}{9}+\frac{1}{2}-2\frac{2}{3}=$

(5) $\frac{1}{12}+9\frac{1}{6}-\frac{7}{18}=$

(6) $12\frac{6}{7}-6\frac{1}{2}+\frac{3}{4}=$

(7) $2\frac{2}{3}+\frac{3}{4}-1\frac{5}{6}=$

(8) $1\frac{2}{3}+1\frac{2}{5}-1\frac{1}{6}=$

(9) $2\frac{1}{6}+\frac{5}{8}-1\frac{7}{12}=$

(10) $3\frac{3}{8}-1\frac{7}{9}+\frac{2}{3}=$

You are ready to try this with word problems.

13 Word Problems with Fractions

Level ★★

Date / /

Name

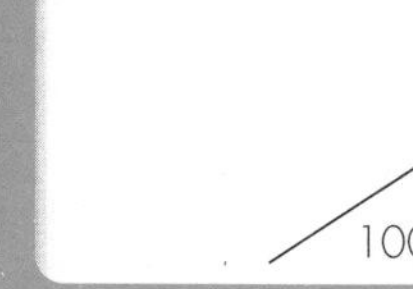

Score /100

■ The Answer Key is on page 90.

1 **Answer each word problem. Write the question as an expression first, and then calculate.** 10 points per question

(1) James pours $\frac{2}{5}$ liter of olive oil from one bottle into another bottle that already has $\frac{1}{10}$ liter of olive oil. How much oil does he have in all?

$$\frac{2}{5}+\frac{1}{10}=$$

〈**Ans.**〉

(2) Sarah used $\frac{1}{9}$ meter of ribbon, and then she used $\frac{1}{3}$ meter more. How much ribbon did she use in all?

〈**Ans.**〉

(3) Otis ran $\frac{1}{2}$ kilometer. His teammate Joris ran $\frac{2}{3}$ kilometer. Then Otis ran another $\frac{1}{6}$ kilometer. How far did they run in all?

$$\frac{1}{2}+\frac{2}{3}+\frac{1}{6}=$$

〈**Ans.**〉

(4) Pat drank $1\frac{1}{3}$ cups of milk. His little sister drank $\frac{1}{2}$ cup of milk. Their dad drank $2\frac{5}{6}$ cups of milk. How much did they drink in all?

〈**Ans.**〉

2 Answer each word problem. Write the question as an expression first, and then calculate.

10 points per question

(1) Mrs. Rubino had $3\frac{3}{4}$ pounds of cherries. She used $1\frac{2}{5}$ pounds of cherries for a pie. How many pounds of cherries does she have left?

$$3\frac{3}{4}-1\frac{2}{5}=3\frac{\square}{20}-1\frac{\square}{20}=$$

〈Ans.〉

(2) There are $2\frac{8}{9}$ cups of flour and $1\frac{1}{3}$ cups of sugar. How much more flour is there than sugar?

〈Ans.〉

(3) Juan has $8\frac{4}{15}$ ounces of salt. He uses $5\frac{7}{10}$ ounces for his first science experiment and $1\frac{1}{5}$ ounces for his second experiment. How much salt does he have left?

〈Ans.〉

(4) The cafeteria has $15\frac{3}{4}$ gallons of soup. They sell $8\frac{7}{12}$ gallons to students and $2\frac{1}{3}$ gallons to teachers. How much soup is left?

〈Ans.〉

3 Answer each word problem. Write the question as an expression first, and then calculate.

10 points per question

(1) Jennie has $1\frac{7}{15}$ pounds of flour in a bowl. She takes $\frac{4}{5}$ pound out, but then adds $\frac{1}{3}$ pound back in. How many pounds of flour are in the bowl now?

〈Ans.〉

(2) Frankie wrote $4\frac{1}{6}$ pages of a story. He didn't like some of it, so he erased $\frac{5}{12}$ page and wrote $\frac{1}{2}$ page more. How many pages long is his story now?

〈Ans.〉

Great job! You are ready for the next step.

Multiplication of Fractions

Level 

Date / /

Name

Score /100

■ The Answer Key is on page 90.

1 **Multiply.**

5 points per question

Example $\frac{8}{9} \times \frac{1}{3} = \frac{8}{27}$

(1) $\frac{1}{4} \times \frac{1}{3} = \frac{\square}{12}$

(2) $\frac{6}{7} \times \frac{3}{5} =$

(3) $\frac{4}{9} \times \frac{4}{7} =$

(4) $\frac{2}{5} \times \frac{4}{5} =$

(5) $\frac{7}{8} \times \frac{3}{4} =$

(6) $\frac{4}{7} \times \frac{5}{9} =$

(7) $\frac{2}{3} \times \frac{7}{9} =$

(8) $\frac{11}{13} \times \frac{1}{2} =$

(9) $\frac{11}{12} \times \frac{1}{4} =$

(10) $\frac{5}{7} \times \frac{5}{12} =$

2 Reduce as you multiply.

5 points per question

Example $\frac{3}{7} \times \frac{4}{9} = \frac{\overset{1}{\cancel{3}}}{7} \times \frac{4}{\underset{3}{\cancel{9}}} = \frac{4}{21}$

(1) $\frac{1}{2} \times \frac{4}{5} = \frac{1}{\underset{1}{\cancel{2}}} \times \frac{\overset{2}{\cancel{4}}}{5} =$

(2) $\frac{10}{11} \times \frac{22}{25} = \frac{\overset{\square}{\cancel{10}}}{\underset{\square}{\cancel{11}}} \times \frac{\overset{\square}{\cancel{22}}}{\underset{\square}{\cancel{25}}} = \frac{\square}{\square}$

(3) $\frac{5}{9} \times \frac{4}{15} =$

(4) $\frac{7}{11} \times \frac{5}{28} =$

(5) $\frac{9}{15} \times \frac{3}{10} =$

(6) $\frac{13}{15} \times \frac{25}{26} =$

(7) $\frac{15}{17} \times \frac{34}{45} =$

(8) $\frac{14}{15} \times \frac{45}{46} =$

(9) $\frac{6}{20} \times \frac{5}{22} =$

(10) $\frac{9}{10} \times \frac{25}{36} =$

Reducing as you multiply will make the problem easier! You won't have to reduce your answer.

Outstanding effort!

Multiplication of Fractions

Level  

Date / / Name

Score /100

■ The Answer Key is on page 90.

1 Multiply.

5 points per question

Example $2\frac{4}{5} \times 1\frac{7}{8} = \frac{\overset{7}{\cancel{14}}}{\underset{1}{\cancel{5}}} \times \frac{\overset{3}{\cancel{15}}}{\underset{4}{\cancel{8}}} = \frac{21}{4} = 5\frac{1}{4}$

Always make your answer a proper fraction.

(1) $1\frac{2}{3} \times 2\frac{1}{4} = \frac{5}{\underset{\square}{\cancel{3}}} \times \frac{\overset{\square}{\cancel{9}}}{4} =$

(2) $2\frac{1}{4} \times 2\frac{2}{13} = \frac{9}{4} \times \frac{28}{13} =$

(3) $7\frac{1}{2} \times 2\frac{2}{5} =$

(4) $3\frac{3}{7} \times 2\frac{7}{12} =$

(5) $4\frac{7}{8} \times 1\frac{1}{13} =$

(6) $10\frac{1}{2} \times 2\frac{2}{7} =$

(7) $5\frac{5}{6} \times 2\frac{2}{5} =$

(8) $4\frac{1}{4} \times 2\frac{8}{17} =$

(9) $12\frac{1}{2} \times 2\frac{4}{5} =$

(10) $5\frac{4}{7} \times 1\frac{8}{13} =$

2 Multiply.

5 points per question

Examples

$$\frac{3}{4}\times\frac{2}{5}\times\frac{5}{11}=\frac{3}{\underset{2}{\cancel{4}}}\times\frac{\overset{1}{\cancel{2}}}{\underset{1}{\cancel{5}}}\times\frac{\overset{1}{\cancel{5}}}{11}=\frac{3}{22}$$

$$1\frac{1}{2}\times4\frac{2}{3}\times2\frac{5}{7}=\frac{\overset{1}{\cancel{3}}}{\underset{1}{\cancel{2}}}\times\frac{\overset{1}{\cancel{\overset{7}{\cancel{14}}}}}{\underset{1}{\cancel{3}}}\times\frac{19}{\underset{1}{\cancel{7}}}=\frac{19}{1}=19$$

(1) $\frac{3}{4}\times\frac{2}{5}\times\frac{5}{9}=$

(2) $\frac{1}{4}\times\frac{8}{9}\times\frac{3}{5}=$

(3) $1\frac{1}{5}\times\frac{3}{4}\times\frac{8}{9}=$

(4) $5\times2\frac{3}{5}\times1\frac{1}{2}=$

(5) $6\times1\frac{2}{3}\times2\frac{3}{4}=$

(6) $2\frac{1}{4}\times\frac{5}{9}\times3\frac{2}{5}=$

(7) $\frac{2}{3}\times5\frac{2}{5}\times3\frac{3}{4}=$

(8) $\frac{2}{13}\times1\frac{6}{7}\times2\frac{4}{5}=$

(9) $12\times1\frac{1}{8}\times1\frac{2}{3}=$

(10) $\frac{4}{11}\times\frac{11}{12}\times1\frac{3}{10}=$

Your math skills are multiplying!

16 Division of Fractions

Level ★★

Date / / Name

Score /100

■ The Answer Key is on page 91.

Divide.

5 points per question

Example $\frac{1}{4} \div \frac{1}{3} = \frac{1}{4} \times \frac{3}{1} = \frac{3}{4}$

When dividing fractions, flip the second fraction and multiply.

(1) $\frac{8}{9} \div \frac{1}{2} = \frac{8}{9} \times \frac{\square}{1} =$

(2) $\frac{4}{5} \div \frac{3}{7} = \frac{4}{5} \times \frac{\square}{\square} =$

(3) $\frac{1}{5} \div \frac{4}{9} =$

(4) $\frac{7}{9} \div \frac{2}{5} =$

(5) $\frac{2}{7} \div \frac{5}{8} =$

(6) $\frac{3}{7} \div \frac{5}{6} =$

(7) $\frac{3}{8} \div \frac{2}{7} =$

(8) $\frac{3}{5} \div \frac{7}{9} =$

(9) $\frac{5}{7} \div \frac{2}{9} =$

(10) $\frac{3}{13} \div \frac{1}{12} =$

Don't forget to make your answer a proper fraction.

2 Divide.

5 points per question

(1) $\frac{6}{7} \div \frac{3}{4} = \frac{\overset{2}{\cancel{6}}}{7} \times \frac{4}{\underset{1}{\cancel{3}}} =$

(2) $\frac{3}{5} \div \frac{9}{10} =$

(3) $\frac{5}{6} \div \frac{3}{4} =$

(4) $\frac{5}{8} \div \frac{7}{8} =$

(5) $\frac{2}{5} \div \frac{14}{15} =$

(6) $\frac{3}{8} \div \frac{1}{4} =$

(7) $\frac{3}{4} \div \frac{5}{8} =$

(8) $\frac{6}{7} \div \frac{4}{5} =$

(9) $\frac{7}{8} \div \frac{5}{6} =$

(10) $\frac{8}{9} \div \frac{10}{11} =$

Don't forget to reduce as you compute!

You can face any challenge!

17 Division of Fractions

Level ★★

Date / / Name

Score /100

■ The Answer Key is on page 91.

1 **Divide.**

5 points per question

(1) $\frac{1}{7} \div 4 = \frac{1}{7} \div \frac{4}{1}$

$= \frac{1}{7} \times \frac{1}{\square}$

$=$

(2) $\frac{6}{7} \div 7 = \frac{6}{7} \div \frac{\square}{1}$

$=$

(3) $\frac{3}{10} \div 4 =$

(4) $5 \div \frac{1}{4} = \frac{\square}{1} \div \frac{1}{4}$

$=$

(5) $4 \div \frac{7}{8} =$

Before dividing, change integers into fractions and change mixed numbers into improper fractions.

2 **Divide.**

5 points per question

(1) $\frac{5}{8} \div \frac{3}{4} =$

(2) $\frac{4}{9} \div 1\frac{5}{6} =$

(3) $1\frac{2}{7} \div \frac{9}{14} =$

(4) $1\frac{5}{9} \div 1\frac{1}{6} =$

(5) $2\frac{1}{3} \div 1\frac{5}{9} =$

Reduce as you compute!

3 Divide.

5 points per question

(1) $\frac{2}{3} \div 6 \div \frac{3}{4} = \frac{2}{3} \div \frac{6}{1} \div \frac{3}{4}$

$= \frac{2}{3} \times \frac{1}{6} \times \frac{4}{3} =$

(2) $\frac{4}{5} \div 1\frac{2}{3} \div 1\frac{1}{2} = \frac{4}{5} \div \frac{\square}{3} \div \frac{3}{\square}$

$=$

(3) $\frac{3}{4} \div 5 \div 1\frac{1}{5} =$

(4) $\frac{7}{24} \div 1\frac{7}{8} \div 3\frac{1}{2} =$

(5) $6 \div 1\frac{2}{3} \div 2\frac{4}{5} =$

(6) $11 \div 2\frac{2}{3} \div 1\frac{4}{7} =$

4 Multiply and divide.

5 points per question

(1) $\frac{5}{6} \times \frac{3}{4} \div \frac{5}{8} = \frac{5}{6} \times \frac{3}{4} \times \frac{8}{5} =$

(2) $\frac{7}{8} \div \frac{1}{2} \times \frac{4}{21} =$

(3) $\frac{2}{9} \times \frac{6}{7} \div \frac{5}{14} =$

(4) $\frac{4}{5} \div \frac{8}{15} \times 1\frac{1}{3} =$

18 Word Problems with Fractions

Level ★★

Date / /

Name

Score /100

■ The Answer Key is on page 91.

1 Answer each word problem. Write the question as an expression first, and then calculate.

10 points per question

(1) Jacob loves to swim. If he swims $1\frac{1}{4}$ miles every day for 6 days, how far will he have swum in all?

〈Ans.〉

(2) The grocer filled 12 bags with cans of food to donate to a local shelter. If each bag weighs $2\frac{4}{9}$ pounds, how much do all the bags weigh together?

〈Ans.〉

(3) Jaolen's new toy car can move $4\frac{1}{4}$ feet in $\frac{2}{3}$ second. How far could it go in 6 seconds?

〈Ans.〉

(4) A recipe calls for 3 cans of condensed milk. Each can holds $\frac{1}{2}$ cup of condensed milk. If Alice wants to cut the recipe in half, how much condensed milk will she need?

〈Ans.〉

2 Answer each word problem.

15 points per question

(1) Jenni loves to crochet. She buys $\frac{5}{7}$ yard of yarn for \$2. How much yarn would she get for \$1?

$\frac{5}{7} \div 2 =$

〈Ans.〉

(2) Mrs. Collins teaches pottery. If she divides $\frac{6}{7}$ pound of clay between 3 students, how much clay does each student get?

〈Ans.〉

(3) 12 liters of water flow out of the faucet every $\frac{3}{5}$ minute. How much water flows out each minute?

〈Ans.〉

(4) The teacher gives each pair of students $1\frac{14}{15}$ liters to water their plants in the garden. If the pairs of students share the water evenly, how much water does each student get?

〈Ans.〉

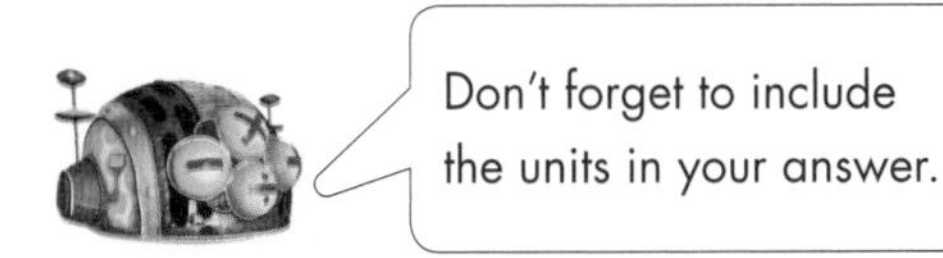

Don't forget to include
the units in your answer.

Look as you go!

19 Fraction Review

Date / /

Name

Level

Score

/100

■ The Answer Key is on page 91.

1 Calculate.

5 points per question

(1) $\frac{5}{7}+\frac{4}{5}=$

(2) $1\frac{3}{10}+2\frac{1}{7}+\frac{2}{5}=$

(3) $5\frac{1}{3}-2\frac{3}{7}=$

(4) $2\frac{11}{14}-1\frac{1}{7}-\frac{1}{2}=$

(5) $5\frac{3}{4}-2\frac{7}{12}+1\frac{1}{3}=$

(6) $5\frac{1}{9}-4\frac{1}{3}+2\frac{2}{5}=$

2 Calculate.

5 points per question

(1) $3\frac{3}{7}\times2\frac{7}{12}=$

(2) $6\times1\frac{1}{4}\times1\frac{2}{3}=$

(3) $1\frac{2}{7}\div1\frac{9}{14}=$

(4) $1\frac{5}{6}\div\frac{1}{2}\div12=$

(5) $11\times2\frac{2}{3}\div1\frac{4}{7}=$

(6) $\frac{7}{24}\times1\frac{7}{8}\div3\frac{1}{2}=$

3 **Answer each word problem. Write the question as an expression first, and then calculate.** 10 points per question

(1) A meatloaf recipe calls for $1\frac{3}{4}$ pounds of ground beef, $1\frac{1}{3}$ pounds of onions, and $1\frac{1}{6}$ pounds of ground sausage. How much do these ingredients weigh in all?

〈Ans.〉

(2) Juliette adds $5\frac{1}{9}$ cups of nuts to $1\frac{1}{3}$ cups of pretzels. She puts $1\frac{5}{18}$ cups of the trail mix in her cupboard and takes the rest to a party. How much trail mix does she take to the party?

〈Ans.〉

4 **Answer each word problem. Write the question as an expression first, and then calculate.** 10 points per question

(1) Sophie jogs $4\frac{1}{4}$ miles in $\frac{7}{8}$ hour. How far would she run in $4\frac{1}{2}$ hours if she kept the same pace?

〈Ans.〉

(2) Lana's toy boat can sail $7\frac{1}{2}$ yards in $4\frac{1}{6}$ minutes. How far would it sail in $8\frac{3}{4}$ minutes if it kept the same pace?

〈Ans.〉

20 Place Value Review

Level ★★

Date / / Name

Score /100

■ The Answer Key is on page 91.

Don't forget!

Whole numbers like 0, 1, 2, and 3 are called **integers**.

Numbers like 0.1, 0.5, and 2.3 are called **decimals**, and the "." is called the **decimal point**.

The places to the right of the decimal point are called **tenths, hundredths, thousandths**, and so on.

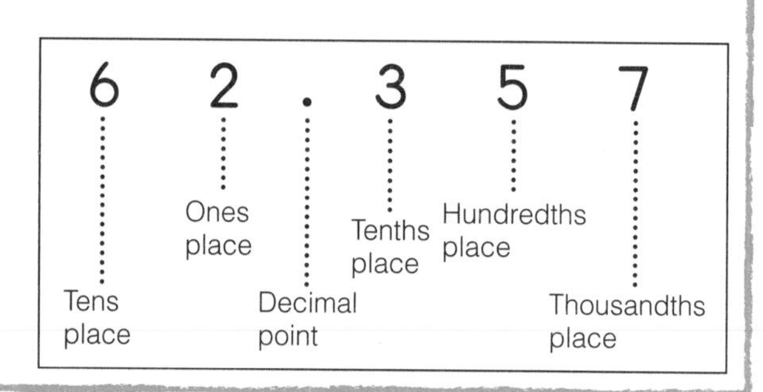

1 **Write the appropriate number in each box.** 4 points per question

(1) The number in the ones place of 4.63 is ☐ and in the tenths place is ☐.

(2) The number in the tenths place of 6.24 is ☐ and in the hundredths place is ☐.

(3) The number in the ones place of 8.712 is ☐ and in the thousandths place is ☐.

(4) The number in the tenths place of 42.903 is ☐ and in the thousandths place is ☐.

(5) The number in the tens place of 76.42 is ☐ and in the tenths place is ☐.

2 **Write the appropriate number in each box.** 5 points per question

(1)

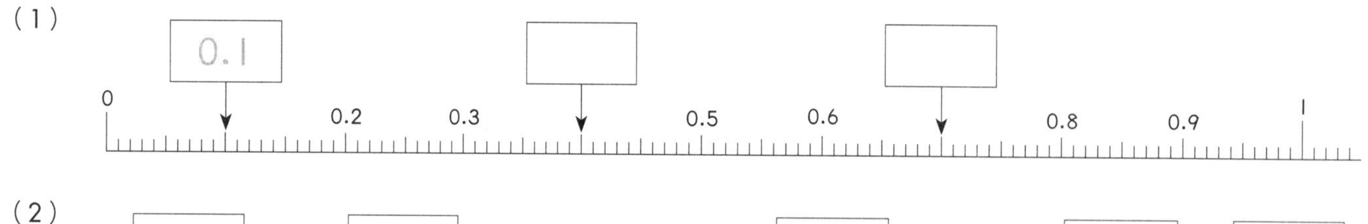

(2)

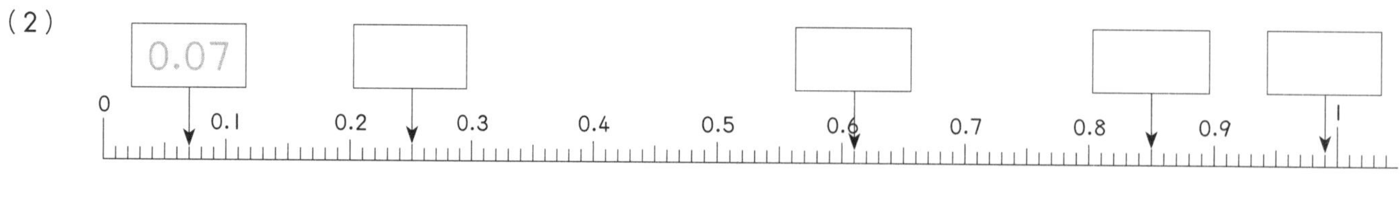

(3)

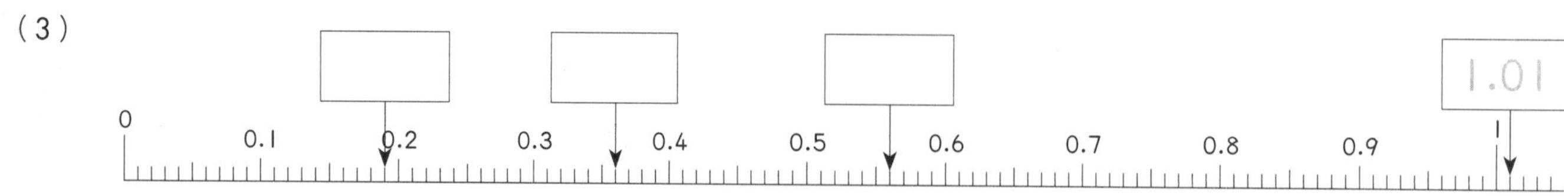

Don't forget!

A **round number** is made by adjusting a number up or down in order to simplify the number for easier calculations. If the number in the next place is 5 or higher, round up. If it is 4 or lower, round down.

Example

Round to the nearest tenths place.

4.6724 ⟶ 4.7

(round up)

Example

Round to the nearest thousandths place.

76.49232 ⟶ 76.492

(round down)

3 **Round each number to the nearest tenths place.** 5 points per question

(1) 6.72 (   ) (2) 8.473 (   ) (3) 0.1574 (   )

4 **Round each number to the nearest hundredths place.** 5 points per question

(1) 8.921 (   ) (2) 4.445 (   ) (3) 0.1119 (   )

5 **Round each number to the nearest thousandths place.** 5 points per question

(1) 0.1574 (   ) (2) 0.1119 (   ) (3) 23.78263 (   )

6 **A scientist is measuring the wing size of bees. Round each length to the nearest thousandths place.** 5 points per question

(1) Bee A's wings measured

0.0146 m (   ) m

(2) Bee B's wings measured

0.0152 m (   ) m

(3) Bee C's wings measured

0.0137 m (   ) m

(4) Bee D's wings measured

0.0163 m (   ) m

21 Decimals as Fractions

Date / /　Name

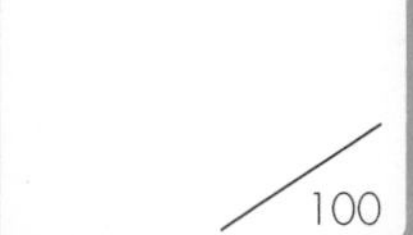

■ The Answer Key is on page 91.

1 **Rewrite each decimal as a fraction. Then reduce.**　4 points per question

Examples

$0.2=\frac{2}{10}=\frac{1}{5}$ (Tenths place)　$0.02=\frac{2}{100}=\frac{1}{50}$ (Hundredths place)　$0.002=\frac{2}{1000}=\frac{1}{500}$ (Thousandths place)

(1) $0.5=\frac{\square}{\square}=\frac{\square}{\square}$

(2) $0.8=\frac{\square}{\square}=\frac{\square}{\square}$

(3) $0.25=\frac{\square}{\square}=\frac{\square}{\square}$

(4) $0.35=\frac{\square}{\square}=\frac{\square}{\square}$

(5) $0.05=\frac{\square}{\square}=\frac{\square}{\square}$

(6) $0.18=\frac{\square}{\square}=\frac{\square}{\square}$

(7) $0.08=\frac{\square}{\square}=\frac{\square}{\square}$

(8) $0.005=\frac{\square}{\square}=\frac{\square}{\square}$

(9) $0.825=\frac{\square}{\square}=\frac{\square}{\square}$

(10) $0.404=\frac{\square}{\square}=\frac{\square}{\square}$

2 Rewrite each decimal as a mixed number. Then reduce.

5 points per question

Examples

$1.4 = 1\frac{4}{10} = 1\frac{2}{5}$ $1.04 = 1\frac{4}{100} = 1\frac{1}{25}$ $1.004 = 1\frac{4}{1000} = 1\frac{1}{250}$

(1) $3.5 =$

(2) $5.75 =$

(3) $2.6 =$

(4) $2.68 =$

(5) $10.55 =$

(6) $9.005 =$

3 Rewrite each decimal as an improper fraction. Then reduce.

5 points per question

Example $3.2 = \frac{32}{10} = \frac{16}{5}$

(1) $2.5 =$

(2) $3.4 =$

(3) $1.12 =$

(4) $2.08 =$

(5) $1.625 =$

(6) $2.008 =$

You can do anything with decimals!

22 Fractions as Decimals

Level 

Date / /

Name

Score /100

■ The Answer Key is on page 92.

1 **Rewrite each fraction as a decimal.**

5 points per question

Examples

$\frac{2}{5} \longrightarrow$

$$\begin{array}{r} 0.4 \\ 5\overline{)2.0} \\ \underline{2\,0} \\ 0 \end{array}$$

$\frac{2}{5} = 0.4$

$\frac{8}{5} \longrightarrow$

$$\begin{array}{r} 1.6 \\ 5\overline{)8.0} \\ \underline{5} \\ 3\,0 \\ \underline{3\,0} \\ 0 \end{array}$$

$\frac{8}{5} = 1.6$

(1) $\frac{1}{4} =$

(2) $\frac{3}{4} =$

(3) $\frac{4}{5} =$

(4) $\frac{7}{5} =$

(5) $\frac{3}{50} =$

(6) $\frac{3}{8} =$

(7) $\frac{33}{4} =$

(8) $\frac{31}{4} =$

2 Rewrite each mixed fraction as a decimal.

6 points per question

Example

$$1\frac{3}{5} = 1.6$$

$$\begin{array}{r} 0.6 \\ 5\overline{)\,3.0} \\ 3\,0 \\ \hline 0 \end{array}$$

(1) $2\frac{1}{4} =$

(2) $7\frac{4}{5} =$

(3) $5\frac{3}{8} =$

(4) $4\frac{3}{25} =$

(5) $3\frac{31}{50} =$

(6) $10\frac{1}{5} =$

(7) $5\frac{12}{25} =$

(8) $2\frac{21}{35} =$

(9) $7\frac{7}{8} =$

(10) $2\frac{5}{8} =$

23 Percents

Level ☆☆

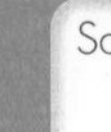

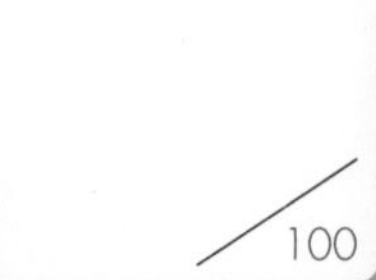

Date / / Name

Score /100

■ The Answer Key is on page 92.

1 **Rewrite each decimal as a fraction and a percent.** 5 points per question

Examples

$0.02 = \frac{2}{100} = 2\%$ (Hundredths place)

$0.20 = \frac{20}{100} = 20\%$ (Hundredths place)

$2.00 = \frac{200}{100} = 200\%$ (Hundredths place)

(1) $0.45 = \frac{\square}{100} =$

(2) $0.35 = \frac{}{100} =$

(3) $0.4 = \frac{}{100} =$

(4) $0.04 = \frac{}{100} =$

(5) $0.004 = \frac{4}{1000} = 0.4\%$

(6) $0.123 = \frac{123}{1000} = 12.3\%$

(7) $0.971 = \frac{}{1000} =$

(8) $0.979 = \frac{}{1000} =$

(9) $0.4805 = \frac{}{10000} =$

(10) $0.0003 = \frac{}{10000} =$

Move the decimal point two places to the right to quickly change a decimal to a percent.

2 Rewrite each number as a percent.

5 points per question

(1) 1 = 100 %

(2) 1.45 =

(3) 1.04 =

(4) 1.045 =

(5) 10.47 =

(6) 1.005 = 100.5 %

(7) 1.012 =

(8) 1.103 =

(9) 2 =

(10) 2.02 =

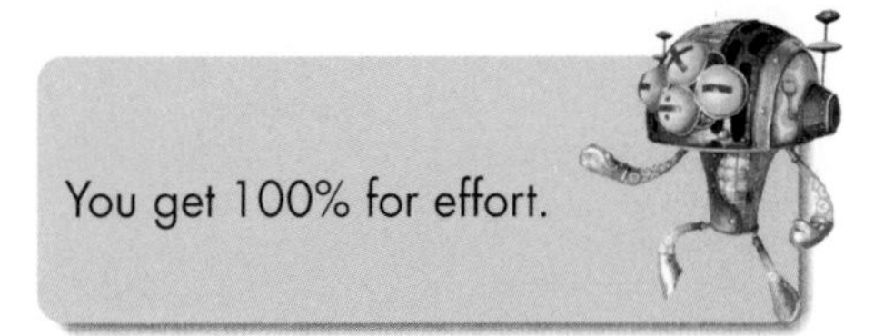

24 Percents

Date / /

Name

Level

Score

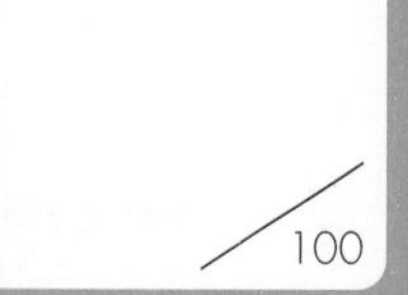

/100

■ The Answer Key is on page 92.

1 **Rewrite each percent as a fraction and a decimal.**

5 points per question

(1) $45\% = \frac{\square}{100} =$

(6) $26.4\% = \frac{\square}{\square} =$

(2) $55\% = \frac{\square}{100} =$

(7) $14.9\% = \frac{\square}{\square} =$

(3) $70\% = \frac{\square}{\square} =$

(8) $0.8\% = \frac{\square}{1000} =$

(4) $87\% = \frac{\square}{\square} =$

(9) $67.02\% = \frac{\square}{\square} =$

(5) $6\% = \frac{\square}{\square} =$

(10) $0.09\% = \frac{\square}{\square} =$

Move the decimal point two places to the left to quickly change a percent into a decimal.

2 Rewrite each percent as a number.

5 points per question

(1) 100 % =

(2) 145 % =

(3) 108 % =

(4) 117 % =

(5) 110 % =

(6) 1,000 % =

(7) 1,100 % =

(8) 1,110 % =

(9) 1,111 % =

(10) 1,011 % =

Percents are challenging, but you can do it!

25 Percents

Level ★★★

Date / /

Name

Score /100

■ The Answer Key is on page 92.

1 **Rewrite each percent as a decimal and a fraction. Then reduce each fraction.**

4 points per question

(1) $25\% = 0.25 = \frac{25}{100} = \frac{1}{4}$

(2) $75\% =$

(3) $60\% =$

(4) $15\% =$

(5) $26\% =$

(6) $150\% =$

(7) $125\% =$

(8) $105\% =$

(9) $10.5\% =$

(10) $100.5\% =$

2 **Rewrite each fraction as a decimal and a percent.**

5 points per question

(1) $\frac{1}{2} = 0.5 = 50\%$

(2) $\frac{1}{4} = \quad =$

(3) $\frac{2}{5} = \quad =$

(4) $\frac{7}{20} = \quad =$

(5) $1\frac{4}{5} = \quad =$

(6) $1\frac{3}{4} = \quad =$

3 Answer each word problem.

5 points per question

(1) Aunt Miriam buys a pen for \$0.82. What percent of a dollar was spent on the pen?

〈Ans.〉

(2) In Teaneck Village, it rained 1.47 times the usual amount during the season. What percentage of the usual amount of rain did the village receive?

〈Ans.〉

4 Answer each word problem.

5 points per question

(1) Yen sells ice cream. His sales increased by 8,250 % since winter. Represent his sales increase as an integer.

〈Ans.〉

(2) Lana adds some tomato plants to her vegetable garden. This increases the size of her garden by 27 %. Represent the size of growth as an integer.

〈Ans.〉

5 Answer each word problem.

5 points per question

(1) Derik pumps gas into his truck so it is $\frac{5}{8}$ full. What percent of his gas tank is full?

〈Ans.〉

(2) A jar of jelly beans is $\frac{1}{5}$ full. What percent of the jar is full?

〈Ans.〉

You've mastered percents!

26 Decimals and Fractions

Level ★★★

Date / /

Name

Score /100

■ The Answer Key is on page 93.

1 **Calculate.**

5 points per question

Convert each decimal into a fraction first to calculate.

(1) $\frac{1}{6}+0.5=$

(2) $0.45+\frac{1}{5}=$

(3) $0.85+\frac{3}{5}=$

(4) $1\frac{5}{6}+0.5=$

(5) $1\frac{1}{8}+0.6=$

(6) $0.75-\frac{1}{3}=$

(7) $\frac{17}{20}-0.55=$

(8) $\frac{7}{12}-0.35=$

(9) $2.75-\frac{2}{3}=$

(10) $4\frac{7}{10}-1.35=$

2 Calculate.

5 points per question

(1) $\frac{1}{4}+0.25+\frac{2}{5}=$

(2) $\frac{5}{6}+0.4+\frac{2}{3}=$

(3) $0.25+\frac{1}{8}+0.6=$

(4) $6\frac{17}{20}-1.4-\frac{3}{10}=$

(5) $7\frac{1}{4}-1.2-3\frac{1}{5}=$

(6) $5\frac{9}{10}-2.3-1\frac{13}{20}=$

(7) $\frac{9}{10}+0.45-\frac{11}{20}=$

(8) $\frac{5}{6}-0.25+\frac{3}{8}=$

(9) $0.875-\frac{3}{4}+1\frac{2}{3}=$

(10) $0.75+\frac{4}{9}-\frac{5}{6}=$

27 Decimals and Fractions

Date / /

Name

Score

■ The Answer Key is on page 93.

1 **Calculate.**

5 points per question

(1) $\frac{1}{6} \times 0.5 =$

(2) $\frac{5}{6} \times 0.75 =$

(3) $\frac{5}{12} \times 0.08 =$

(4) $2\frac{2}{3} \times 0.9 =$

(5) $0.04 \times 1\frac{2}{3} =$

(6) $0.45 \div \frac{1}{4} =$

(7) $0.55 \div \frac{9}{10} =$

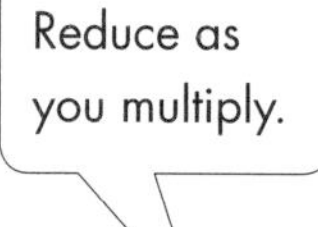

(8) $0.325 \div \frac{1}{6} =$

(9) $1\frac{1}{12} \div 0.13 =$

(10) $0.625 \div 2\frac{1}{12} =$

Convert each decimal into a fraction to calculate.

2 Calculate.

5 points per question

(1) $\frac{5}{6}\times\frac{2}{3}\times0.9=$

(2) $\frac{5}{8}\div\frac{8}{9}\times0.8=$

(3) $\frac{5}{7}\times0.4\div\frac{3}{7}=$

(4) $0.6\times\frac{2}{3}\div0.42=$

(5) $\frac{5}{7}\times0.25\div\frac{2}{7}=$

(6) $\frac{3}{4}\div\frac{2}{3}\div0.2=$

(7) $1.8\div\frac{5}{8}\times\frac{8}{9}=$

(8) $2\frac{5}{7}\times1.4\div1\frac{3}{7}=$

(9) $2\frac{1}{3}\times4\times1.2=$

(10) $1\frac{5}{12}\div1.25\div1\frac{2}{3}=$

28 Word Problems with Decimals and Fractions

■ The Answer Key is on page 93.

1 **Answer each word problem. Write the question as an expression first, and then calculate.** 10 points per question

(1) Spencer pours 0.15 liter of water out of a full jug that holds $\frac{9}{10}$ liter. Then he adds $\frac{3}{100}$ liter back. How much water is in the jug?

〈Ans.〉 ______________________

(2) Sheila knits $6\frac{9}{10}$ feet of a blanket. She unravels 2.15 feet off of it and knits $\frac{1}{6}$ foot back onto it. How long is the blanket?

〈Ans.〉 ______________________

2 **Answer each word problem. Write the question as an expression first, and then calculate.** 10 points per question

(1) Christopher has 4.5 pieces of wood. Each full piece is $1\frac{8}{15}$ yards long. How long is the wood if Christopher glues the pieces together?

〈Ans.〉 ______________________

(2) Gene has $3\frac{1}{3}$ pieces of ribbon. Each full piece is 2.15 meters long. How long is the ribbon if Gene glues the pieces together?

〈Ans.〉 ______________________

3 **Answer each word problem. Write the question as an expression first, and then calculate.** 15 points per question

(1) Matthew has $3\frac{1}{3}$ pieces of copper pipe. Each full piece is 2.5 meters long. If Matthew welds the pieces together and then divides the new pipe into $\frac{5}{6}$ meter pieces, how many pieces will he have?

⟨Ans.⟩

(2) Dominic has 6.75 pieces of wood. Each full piece is 1.2 yards long. If he glues the pieces together and then cuts the new wood every $\frac{9}{10}$ yards, how many pieces of wood will he have?

⟨Ans.⟩

4 **Answer each word problem. Write the question as an expression first, and then calculate.** 15 points per question

(1) A baker uses 0.6 ton of whole wheat flour each month. $\frac{1}{9}$ of the whole wheat flour is used for muffins. If the baker adds 0.1 ton of all-purpose flour to the muffin mix, how much does the muffin mix weigh?

⟨Ans.⟩

(2) A scientist has a 1.5 liter bottle of water. He pours $\frac{9}{10}$ of the bottle into a beaker and boils it so 0.25 liter of the liquid evaporates. How much water is left in the beaker?

⟨Ans.⟩

You can work through anything!

Exponents

Date / /　　Name

Level ★★

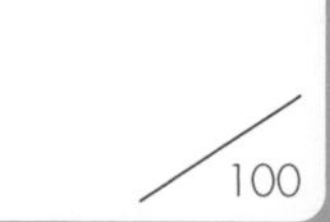

■ The Answer Key is on page 93.

1 Calculate.

5 points per question

Examples

$4^2 = 4 \times 4 = 16$　　4^2 is read as "four squared."

$4^3 = 4 \times 4 \times 4 = 64$　　4^3 is read as "four cubed."

(1) $2^2 = 2 \times 2 =$

(2) $2^3 = 2 \times 2 \times 2 =$

(3) $2^4 = 2 \times 2 \times 2 \times 2 =$

(4) $3^2 =$

(5) $3^3 =$

(6) $3^0 = 1$

(7) $3^5 =$

(8) $4^4 =$

(9) $6^0 =$

(10) $7^3 =$

Any number raised to the power of 0 equals 1.

3^4 can be read as "three to the fourth power." 3^5 can be read as "three to the fifth power."

2 Calculate.

5 points per question

Examples $\left(\frac{2}{3}\right)^2 = \frac{2}{3} \times \frac{2}{3} = \frac{4}{9}$ $\frac{2}{3^2} \times \frac{2}{3 \times 3} = \frac{2}{9}$

(1) $\left(\frac{1}{2}\right)^2 = \frac{1}{2} \times \frac{1}{2} =$

(4) $\frac{1}{4^2} = \frac{1}{4 \times 4} =$

(2) $\left(\frac{1}{2}\right)^3 =$

(5) $\frac{3}{5^3} =$

(3) $\left(\frac{1}{4}\right)^2 =$

(6) $\left(\frac{3}{5}\right)^0 =$

3 Answer each word problem.

10 points per question

(1) The volume of a cube can be calculated by multiplying the width, length, and height, which are all equal lengths. If the width, length, and height of Sunny's block each equal $\frac{3}{4}$ inch, what is the volume of the block?

$\left(\frac{3}{4}\right)^3 =$

⟨**Ans.**⟩ in^3

(2) The cafeteria cook wants to find out the volume of a cube cooler he will fill with ice. If one side of the cube is $\frac{4}{5}$ foot, what is the volume of the cooler?

$\left(\frac{4}{5}\right)^3 =$

⟨**Ans.**⟩ ft^3

30 Exponents

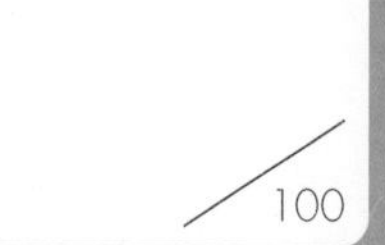

Level ★★

Date / /　Name

Score /100

■The Answer Key is on page 93.

1 Calculate.

4 points per question

(1) $\left(1\frac{1}{3}\right)^2 = \left(\frac{4}{3}\right)^2 =$

(2) $\left(1\frac{1}{3}\right)^3 =$

(3) $\left(1\frac{1}{2}\right)^3 =$

(4) $\left(1\frac{1}{2}\right)^4 =$

(5) $\left(2\frac{1}{2}\right)^2 =$

(6) $\left(1\frac{2}{3}\right)^2 =$

(7) $\left(2\frac{1}{4}\right)^2 =$

(8) $\left(1\frac{2}{3}\right)^3 =$

(9) $\left(2\frac{1}{4}\right)^0 =$

(10) $\left(3\frac{3}{4}\right)^2 =$

Convert mixed numbers to improper fractions before calculating.

Don't forget that any number raised to the power of 0 equals 1.

2 Calculate.

5 points per question

(1) $2^5 \times \frac{1}{2^2} = 2 \times 2 \times 2 \times 2 \times 2 \times \frac{1}{2 \times 2}$

$=$

(2) $3^4 \times \left(\frac{1}{6}\right)^3 = 3 \times 3 \times 3 \times 3 \times \frac{1}{6 \times 6 \times 6}$

$=$

(3) $4^3 \times \left(\frac{1}{4}\right)^2 =$

(4) $5^4 \times \frac{1}{10^3} =$

(5) $8^2 \times \frac{3}{4^2} =$

(6) $9^3 \times \left(\frac{2}{3}\right)^4 =$

3 Calculate.

5 points per question

(1) $\left(1\frac{1}{3}\right)^2 \times \left(\frac{3}{4}\right)^3 = \left(\frac{4}{3}\right)^2 \times \left(\frac{3}{4}\right)^3$

$= \frac{4}{3} \times \frac{4}{3} \times \frac{3}{4} \times \frac{3}{4} \times \frac{3}{4} =$

(2) $\left(1\frac{1}{2}\right)^3 \times 6^2 =$

(3) $8^2 \times \left(2\frac{1}{2}\right)^4 =$

(4) $9^2 \times \left(1\frac{1}{3}\right)^4 =$

(5) $\left(2\frac{1}{2}\right)^4 \times \left(2\frac{2}{5}\right)^2 =$

(6) $\left(1\frac{1}{4}\right)^2 \times \frac{2}{3^3} =$

Convert mixed numbers to improper fractions before calculating.

31 Order of Operations

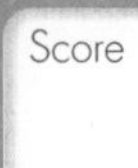

Level ★★

Score /100

Date / /

Name

■ The Answer Key is on page 93.

1 **Calculate. Write the intermediate steps taken to calculate each answer.** 5 points per question

Don't forget!

According to the order of operations,

- **calculate the numbers in parentheses first**
- then calculate from left to right

Example $6-(7-2)=6-5=1$

(1) $7-(8-3)=7-\square=$

(2) $8-(9-5)=$

(3) $8+(10-9)=$

(4) $(5-2)+6=$

(5) $(8-5)+4=$

(6) $(10-2)\times7=8\times7=$

(7) $4\times(2-1)=$

(8) $2\times(8-5)=$

(9) $(15-7)\div2=$

(10) $18\div(8-5)=$

2 Calculate. Write the intermediate steps.

5 points per question

(1) $\frac{5}{7} - \left(\frac{6}{7} - \frac{2}{7}\right) = \frac{5}{7} - \frac{\square}{7} =$

(2) $\frac{1}{2} + \left(5 - \frac{1}{4}\right) = \frac{1}{2} + \left(\frac{\square}{4} - \frac{1}{4}\right)$

$=$

(3) $4 \times \left(\frac{5}{6} - \frac{2}{3}\right) =$

(4) $\left(\frac{4}{5} - \frac{2}{3}\right) \div \frac{1}{3} =$

(5) $(10 - 2) + (13 - 6) =$

(6) $(6 - 4) + \left(1\frac{5}{6} - 1\frac{2}{3}\right) =$

(7) $4 \times (7 \div 2) = \overset{2}{4} \times \frac{7}{\underset{1}{2}} =$

(8) $(9 \div 4) \times \frac{1}{3} =$

(9) $(7 - 3) \times \left(\frac{1}{2} + \frac{1}{4}\right) =$

(10) $\left(\frac{4}{5} + \frac{2}{3}\right) \div \left(\frac{3}{5} - \frac{1}{3}\right) =$

You thought this through!

32 Order of Operations

Level 

Date / / Name

Score 

/100

■ The Answer Key is on page 94.

1 **Calculate. Write the intermediate steps.** 5 points per question

Don't forget!

According to the order of operations,

- **calculate the numbers in parentheses and brackets first (start with the innermost and continue to the outermost)**
- then calculate from left to right

Example $8-[9-(2+3)]=8-[9-5]=8-4=4$

(1) $[4+(8-1)]-9=[4+\square]-9$

$=\square-9$

$=$

(2) $[12-(15-7)]+3=$

(3) $20-[(4+8)-1]=$

(4) $2\times[9-(6-3)]=$

(5) $7\div[2+(6-2)]=$

(6) $[3\div(2-1)]+\frac{5}{7}=[\square\div\square]+\frac{5}{7}$

$=\square+\frac{5}{7}$

$=3\frac{5}{7}$

(7) $9\times\left[\frac{1}{6}+\left(\frac{1}{2}-\frac{1}{3}\right)\right]=$

(8) $\left[\frac{3}{4}+\left(2-\frac{1}{2}\right)\right]+1\frac{5}{7}=$

(9) $\left[\left(\frac{1}{5}+1\frac{1}{10}\right)-\frac{1}{2}\right]\div\frac{1}{5}=$

(10) $\left[(10-8)-1\frac{6}{7}\right]+\left[9-\left(4-\frac{5}{7}\right)\right]$

$=$

2 Calculate. Write the intermediate steps.

5 points per question

(1) $6-\left[\left(\frac{1}{4}+0.75\right)-\frac{3}{4}\right]$

$=6-\left[\left(\frac{1}{4}+\frac{\square}{\square}\right)-\frac{3}{4}\right]=$

(2) $4-\left[\frac{1}{3}+(1.5-0.4)\right]$

$=4-\left[\frac{1}{3}+1.1\right]=4-\left[\frac{1}{3}+1\frac{\square}{\square}\right]$

$=$

(3) $\left[2.6+\left(2\times\frac{1}{2}\right)\right]-2.8$

$=$

(4) $\left[3\times\left(1.75-\frac{3}{4}\right)\right]+3.5$

$=$

(5) $2\times\left[1\frac{1}{2}+(3.3-1.1)\right]$

$=$

(6) $\left[5\frac{5}{6}-\left(2.5-\frac{2}{3}\right)\right]\div\frac{2}{3}$

$=$

(7) $0.5\times\left[16\times\left(2-1\frac{1}{4}\right)\right]$

$=$

(8) $17\div\left[0.4+\left(1\frac{5}{6}-1\frac{2}{3}\right)\right]$

$=$

(9) $\frac{5}{6}\times\left[3\div\left(0.625-\frac{3}{8}\right)\right]$

$=$

(10) $5.4+\left[2+\left(1\frac{1}{4}\div\frac{3}{4}\right)\right]$

$=$

Convert decimals to fractions if necessary to calculate.

Nicely done!

33 Order of Operations

Level ☆☆☆

Date / / Name

Score /100

■ The Answer Key is on page 94.

1 Calculate. Write the intermediate steps.

5 points per question

Don't forget!

According to the order of operations,

- **calculate exponents and numbers in parentheses and brackets first (start with the innermost and continue to the outermost)**
- then calculate from left to right

Example $(5-2)^3\times\frac{1}{9}=3^3\times\frac{1}{9}=\overset{3}{\cancel{27}}\times\frac{1}{\underset{1}{\cancel{9}}}=3$

(1) $3^2-[4+(2+3)]=3^2-[4+\square]$

$=3^2-\square=\square-\square=$

(2) $(18\div9)^3\times(3+1)=$

(3) $[(4+5)+3^2]+3^0=$

(4) $[15-(7+6)]^3-(4+2)$

$=$

(5) $[2+(3^2-4)]\times\frac{5}{7}$

$=[2+(\square-4)]\times\frac{5}{7}$

$=[2+\square]\times\frac{5}{7}=$

(6) $7\div\left[3^2-\left(\frac{1}{2}+\frac{1}{3}\right)\right]$

$=7\div\left[3^2-\left(\frac{3}{6}+\frac{2}{6}\right)\right]=$

(7) $3^2-\left[\frac{1}{4^2}+\left(\frac{1}{2}+\frac{1}{4}\right)\right]=$

(8) $\left[\left(\frac{1}{4^2}+\frac{9}{2^4}\right)\div\frac{1}{2^2}\right]+1=$

2 Calculate. Write the intermediate steps.

6 points per question

(1) $4^2+\left[\frac{1}{6^2}+\left(\frac{1}{2}-\frac{1}{4}\right)\right]$

$=$

(2) $\left(\frac{5}{7}\right)^2+[10+(8-7)^5]$

$=$

(3) $\left[\left(\frac{1}{2}\right)^3+\left(\frac{1}{4}\right)^2\right]\times 4^2$

$=$

(4) $2^3\times\left[\left(\frac{1}{3}\right)^2-\left(\frac{1}{6}\right)^2\right]$

$=$

(5) $\left[\frac{3}{4}-\left(\frac{2}{3}\right)^2\right]\times\left(\frac{2}{3}\right)^2$

$=$

(6) $\left[\left(2\frac{2}{3}\right)^2-\frac{8}{9}\right]-\frac{2}{3^2}$

$=$

(7) $\frac{2^2}{5}+\left[2^3-\left(1\frac{1}{2}\right)^2\right]$

$=$

(8) $\left[2\frac{1}{10}-\left(1\frac{1}{5}\right)^2+2^2\right]\div\left(\frac{3}{10}\right)^2$

$=$

(9) $[12-(5-3)^3]+\left[\left(1\frac{1}{8}\right)^2-\frac{1}{2^3}\right]$

$=$

(10) $\left[\frac{8}{9}\div\left(\frac{8}{9}-\frac{2}{3}\right)^2\right]\times\left(1\frac{1}{3}\right)^2$

$=$

These are tough, but you can do it!

34 Order of Operations

Level ★★★

Date / /

Name

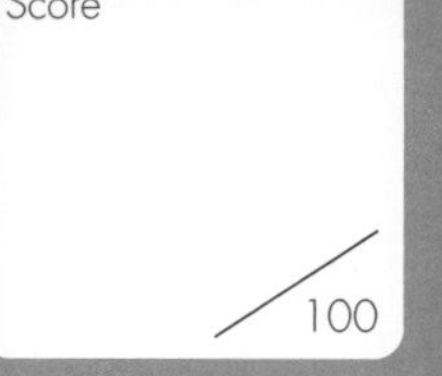

Score /100

■ The Answer Key is on page 94.

1 **Calculate. Write the intermediate steps.**

5 points per question

(1) $3^2-\left(0.5+\frac{1}{4}\right)=$

(2) $2^4-\left(\frac{2}{3}+0.75\right)=$

(3) $0.25+2^3-\frac{1}{2^2}=$

(4) $4^2+\left(0.75-\frac{1}{3^2}\right)=$

(5) $\left[7\div\left(0.25\times\frac{4}{5}\right)\right]-5^2$

$=$

(6) $\left[4^2\times\left(0.1+\frac{2}{5}\right)\right]+1\frac{3}{5}$

$=$

(7) $10^2\div\left[0.2\div\left(\frac{1}{5}\right)^2\right]+2^4$

$=$

(8) $\left[2.6+\left(\frac{7}{15}-\frac{2}{5}\right)\right]\div\frac{1}{3}$

$=$

Convert decimals to fractions whenever necessary.

2 Calculate. Write the intermediate steps.

10 points per question

(1) $\left[\left(1\frac{1}{4}+0.25\right)\times 2^2\right]\div\frac{1}{3}$

$=$

(2) $\left[\left(1\frac{2}{3}+1.75\right)\times 3\right]-\frac{1}{2^2}$

$=$

(3) $\left[\left(\frac{2}{3}+4.125\right)+\left(\frac{1}{2}\right)^2\right]+2^3$

$=$

(4) $\frac{2}{4^2}\times\left[\left(1.5\times 2\frac{1}{3}\right)\div\frac{7}{8}\right]^2$

$=$

(5) $\left(1.375+\frac{9}{2^3}\right)+(18-4^2)^2+2\frac{5}{8}$

$=$

(6) $\left(2.6-1\frac{3}{5}\right)^2+3^3-\left(4.2-3\frac{1}{5}\right)^3$

$=$

Calculate from inside to outside and left to right.

35 Order of Operations

Level ★★★

Date / /

Name

Score /100

■ The Answer Key is on page 95.

1 **Calculate. Write the intermediate steps.**

5 points per question

Don't forget!

According to the order of operations,

- calculate exponents and numbers in parentheses and brackets first
- **perform multiplication and division before addition and subtraction**
- then calculate from left to right

Examples $2+6\times3=2+18=20$ $12\div4-1=3-1=2$

(1) $6-8\div2=6-\square=$

(6) $3\frac{1}{2}-3\div2=$

(2) $4+3\times2=$

(7) $11\div2+4\frac{1}{2}=$

(3) $20-18\div2=$

(8) $6+2\times1\frac{3}{4}=$

(4) $\frac{2}{3}+2\div3=\frac{2}{3}+\frac{\square}{\square}=$

(9) $2\frac{2}{3}-4\div9=$

(5) $8-\frac{2}{7}\times\frac{7}{8}=$

(10) $\frac{5}{6}-\frac{1}{3}\div\frac{1}{2}+1=$

2 Calculate. Write the intermediate steps.

5 points per question

(1) $1\frac{3}{8}\times2-1\frac{1}{4}=$

(6) $2\frac{2}{5}\div1\frac{7}{17}-\frac{11}{15}=$

(2) $3\frac{1}{4}-2\frac{1}{3}\times\frac{3}{4}=$

(7) $3\frac{3}{4}\times\frac{2}{3}-1\frac{3}{4}=$

(3) $2\frac{1}{5}-\frac{2}{3}\times\frac{3}{4}=$

(8) $3-2\frac{1}{4}\div2\frac{5}{8}=$

(4) $4\frac{1}{6}-2\frac{2}{3}\div3\frac{1}{5}=$

(9) $2+1\frac{2}{5}\times3\frac{4}{7}=$

(5) $6\frac{1}{2}+3\frac{2}{5}\times\frac{5}{6}=$

(10) $1\frac{5}{9}\div2\frac{1}{3}-\frac{1}{6}=$

Outstanding effort!

36 Word Problems with Order of Operations

Level ☆☆☆

Date / / Name

Score /100

■ The Answer Key is on page 95.

1 **Answer each word problem. Write the question as an expression first, and then calculate.** 10 points per question

(1) Ricardo baked 15 cupcakes for a party. 16 guests were invited, but 3 people didn't come. If each attending guest had a cupcake, how many cupcakes were left over?

$\square - (\square - \square) =$

⟨Ans.⟩

(2) Matilda promised to make a bracelet for each of her 9 friends. She made the same promise to her 5 cousins. If she makes 8 bracelets on Saturday and 8 more on Sunday, how many bracelets will she have left over?

$(\square + \square) - (\square + \square) =$

⟨Ans.⟩

2 **Answer each word problem. Write the question as an expression first, and then calculate.** 10 points per question

(1) To find the area of a square room, Gordon must square the length of one side. The length of one side of room A is $1\frac{1}{4}$ yards. The length of one side of room B is 6 yards. What is the area of both rooms together?

$(\square)^2 + \square^2 =$

⟨Ans.⟩ yd^2

(2) To find the area of a rectangular room, Gordon must multiply the length by the width. The length of room A is 2 yards and the width is $\frac{1}{2}$ yard. Room B is a square, and the length of one side is 4 yards. Room C is the same size as room B. What is the area of all the rooms together?

$(\square \times \square) + \square^2 + \square^2 =$

⟨Ans.⟩ yd^2

3 Answer each word problem. Write the question as an expression first, and then calculate.

15 points per question

(1) Allison has 4 bags of sugar. Each bag holds $1\frac{5}{8}$ cups. If she combines all of the sugar together in a bowl, and then takes out $1\frac{3}{4}$ cups, how many cups will be in the bowl?

〈Ans.〉

(2) Paul has 5 containers of glass beads. Each container holds $2\frac{1}{4}$ cups. He combines all of the beads together in a bowl and then divides the beads evenly into 4 containers. How many cups of beads is in each container?

〈Ans.〉

4 Answer each word problem. Write the question as an expression first, and then calculate.

15 points per question

(1) Jean-Paul bought 2 crates of apples. Each crate weighed $3\frac{5}{6}$ kilograms. He put $1\frac{1}{2}$ kilograms aside for eating and used the rest to bake pies for the bake sale. How many kilograms of apples did he use for the pies?

〈Ans.〉

(2) Yesterday, Jan read $\frac{1}{6}$ of his book in the morning and $\frac{1}{4}$ of it in the evening. If he does this again today, how much of his book will he have read?

〈Ans.〉

You figured it out!

37 Order of Operations

Level ★★★

Date / /

Name

Score /100

■ The Answer Key is on page 95.

1 **Calculate. Write the intermediate steps.**

5 points per question

(1) $3+[4\times(4-1)]\div 2$

$=3+[\square\times\square]\div 2$

$=3+\square\div 2=3+\square=$

(2) $[(2+2^2)\div 3]+4$

$=$

(3) $[8^2\div(2+2)]\div 4$

$=$

(4) $6+24\div(4+2^2)$

$=$

(5) $[27\div(2+1)^2]+4^0$

$=$

(6) $[10\div(8-3)]\times 3^2$

$=$

(7) $[90\div(8-5)^2]-2^2$

$=$

(8) $[42\div(3+3)]-8^2\div 16+2$

$=$

(9) $[96\div(4^2-4)]+(54-3^2)$

$=$

(10) $[(5+1)^2\div 4]-2+45\div 3^2$

$=$

Always calculate from inside to outside and left to right. Perform multiplication and division before addition and subtraction.

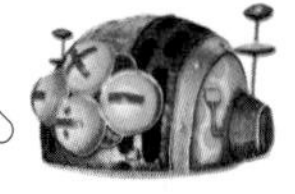

2 Calculate. Write the intermediate steps.

5 points per question

(1) $4^2+[8\times(5-3)]$

$=$

(2) $[1+(2^2-1)]-3$

$=$

(3) $[(8\div2)^2-8]\div2$

$=$

(4) $60\div12+(8^2-5^2)$

$=$

(5) $8-[24\div(4+4^2\div2)]$

$=$

(6) $[10\div(3+2)]\times(8-5)^2$

$=$

(7) $[15\div(13-8)]^3\div9+2$

$=$

(8) $[(5-2)^2-45\div3^2]+1$

$=$

(9) $45\div3^2\times[(7-5)\times2]^2$

$=$

(10) $42\div7-[(6+2)^2\div16]$

$=$

You did it in the correct order!

Order of Operations

Level ★★★

Date / /

Name

Score /100

■ The Answer Key is on page 95.

1 Calculate. Write the intermediate steps.

6 points per question

(1) $\left(\frac{3}{4}\right)^2+(7\times5-3)$

$=\left(\frac{\square}{\square}\times\frac{\square}{\square}\right)+(\square-3)=\frac{\square}{\square}+\square=$

(2) $3\div\left(\frac{2}{5}+2^3\right)+4$

$=$

(3) $\left(\frac{2}{5}+\frac{2}{5}\right)^2+2^5\div8$

$=$

(4) $6\times\left(1\frac{1}{2}-\frac{2}{3}\right)+3^2$

$=$

(5) $\left(\frac{1}{2}\right)^2\times20+1\frac{1}{5}\times5^2$

$=$

(6) $\left(\frac{1}{2}\times1\frac{1}{5}\right)\times(80\div2^2)$

$=$

(7) $\left(\frac{1}{5}+\frac{3}{4}-\frac{3}{10}\right)\times2^2$

$=$

(8) $\left(\frac{1}{4}-\frac{1}{6}\right)\times(5^2-1)$

$=$

(9) $\left(\frac{3}{16}\times8+\frac{1}{2}\right)\div24$

$=$

(10) $\left(\frac{1}{4}+\frac{7}{8}\right)^2+5\div2+2^2$

$=$

2 Calculate. Write the intermediate steps.

5 points per question

(1) $\frac{3}{4^2}+[7\times(5-3)]=\frac{3}{\square\times\square}+[7\times2]$

$=$

(2) $5\div\left[\left(\frac{2}{5}\right)^2\div\frac{4}{15}\right]\div\frac{1}{4}$

$=$

(3) $\left[(3\times2)^2\times\left(1\frac{2}{3}-\frac{1}{2}\right)\right]+3^2$

$=$

(4) $\left[\left(1\frac{1}{2}\right)^3\times4^2\right]\div3^2$

$=$

(5) $\left[\left(1\frac{2}{5}\right)^2-\frac{9}{25}\right]\times5^2\times\frac{1}{2^2}$

$=$

(6) $\left[\left(\frac{3}{4}-\frac{1}{6}\right)^2\div\frac{7}{8}\right]+4^2$

$=$

(7) $\left(2\frac{1}{2}\right)^2\times\left(2+1\frac{3}{5}\right)$

$=$

(8) $\left(7\frac{7}{10}-2\frac{1}{5}+4\frac{3}{4}\right)\div\frac{1}{2^2}$

$=$

Keep up the good work!

Order of Operations

Date / /

Name

Score

■ The Answer Key is on page 95.

1 **Calculate. Write the intermediate steps.** 8 points per question

(1) $\left(1\frac{1}{2}\right)^3 \div \frac{3}{4} + 5.5$

=

(4) $7.6 \div \left(2\frac{1}{5} - \frac{14}{15}\right) + 2^3$

=

(2) $\left(3\frac{4}{5} - 2.8 \times \frac{4}{7}\right) + 0.4$

=

(5) $\left(\frac{2}{3}\right)^2 \times \left(2\frac{7}{8} + 1\frac{2}{5}\right) \div 0.3$

=

(3) $\left(\frac{2}{9}\right)^2 \times \left(1.7 - 1\frac{1}{4}\right)$

=

Convert decimals to fractions in order to calculate.

2 Calculate. Write the intermediate steps.

(1) $1\frac{5}{9}\times\left[\left(\frac{19}{20}-0.65\right)^2\div\frac{3}{20}\right]$

$=$

(2) $\left[3\times\left(1\frac{1}{3}-\frac{1}{4}\right)\div1\frac{1}{12}\right]^3$

$=$

(3) $\left[(3.75\div5)^2+3\frac{3}{8}\right]-2.25$

$=$

(4) $\left[\left(3\frac{3}{4}\div3\right)\times0.8\right]^2+3\frac{1}{8}$

$=$

(5) $\left[\left(2\frac{3}{10}+1\frac{3}{5}\right)\div0.6+6\right]\div5$

$=$

(6) $4\frac{3}{8}\times\frac{3}{5}\div\left[\left(\frac{5}{8}-0.375+5\right)\div3^2\right]$

$=$

This is hard, but you're doing great!

Order of Operations

Date / /

Name

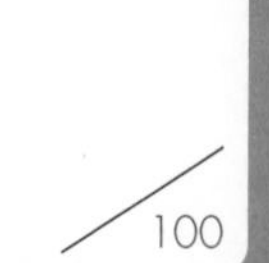

Score

■ The Answer Key is on page 95.

1 Calculate. Write the intermediate steps.

8 points per question

(1) $2\frac{2}{9} \div \left[\left(\frac{17}{20} - 0.55\right)^2 \div 0.54\right]$

$=$

(2) $(0.4)^2 \times \left[\left(\frac{1}{2}\right)^2 \times 2^3 \times 3 \div 4\right]^2$

$=$

(3) $\left(2\frac{1}{2}\right)^2 \times \left[\left(\frac{5}{6}\right)^2 \div 15\right] \div \left(\frac{5}{12}\right)^2$

$=$

(4) $\left[1\frac{3}{5} + 5^3 \div \left(2\frac{1}{2}\right)^2\right] \div 2\frac{2}{5}$

$=$

(5) $\left[\left(3\frac{3}{4} \div 3\right)^2 \times 0.8\right]^2 - \left[1\frac{1}{8} - \left(\frac{3}{4}\right)^2\right]$

$=$

2 **Calculate. Write the intermediate steps.**

10 points per question

(1) $\left[2\times\left(1\frac{1}{3}-0.25\right)\div1\frac{1}{12}\right]^2$

$=$

(2) $\left[\left(3\frac{3}{4}\div5\right)^2+3.375\right]-2\frac{5}{8}$

$=$

(3) $\left(2\frac{1}{2}\right)^2\times\left[\left(\frac{5}{6}\right)^2\times1\frac{1}{15}\right]\div\left(1\frac{2}{3}\right)^2$

$=$

(4) $\left[2\frac{4}{5}-2\frac{2}{9}\times\left(\frac{1}{2}-0.35\right)^2\right]\div5\frac{1}{2}$

$=$

(5) $\left[\left(2.3+1\frac{3}{5}\right)\div\frac{3}{5}-1^2\right]\div5\frac{1}{2}$

$=$

(6) $(6^2-5^2+2)\div\left(3+\frac{1}{2^2}\right)$

$=$

Bravo!

Order of Operations

Level ★★★

Date / /

Name

Score

/100

■ The Answer Key is on page 96.

1 **Calculate. Write the intermediate steps.**

8 points per question

(1) $\left[(3.75 \div 5)^2 + 4\frac{3}{8}\right] - 0.625$

=

(2) $\left[3 \times \left(2\frac{1}{4} - 1.25\right) \div 1\frac{1}{4}\right]^2$

=

(3) $3\frac{2}{3} \div \left[2.7 - 3\frac{1}{8} \times \left(\frac{3}{4} - 0.35\right)^2\right]$

=

(4) $(3^3 - 2 \times 7) \div \left[3 + \left(\frac{1}{2}\right)^2\right]$

=

(5) $3^2 \times \left(4\frac{1}{3} - 3.25\right) \div 1\frac{1}{12}$

=

2 Calculate. Write the intermediate steps.

10 points per question

(1) $(5^2-12)\div\left[2^2-1+\left(\frac{1}{2}\right)^2\right]$

$=$

(2) $\frac{2}{3}\times\left[\left(0.65-\frac{7}{20}\right)^2\div0.66\right]$

$=$

(3) $4\frac{1}{2}\times[(2.25+3)\div3^2]$

$=$

(4) $\frac{1}{3}\div\left[\left(\frac{3}{10}\right)^2\div0.54\right]$

$=$

(5) $[6^2-(4^2-3)]\div\left(2+2\frac{3}{5}\right)$

$=$

(6) $\frac{3}{4^2}+(5^2-12)\div[3+(0.5)^2]$

$=$

You've mastered this!

Review

Level ☆☆

Date / /

Name

Score /100

■The Answer Key is on page 96.

1 Reduce each pair of fractions and compare.

3 points per question

(1) $\frac{7}{70} = \underline{\quad}$, $\frac{9}{30} = \underline{\quad}$

$\underline{\quad} \square \underline{\quad}$

(3) $\frac{3}{18} = \underline{\quad}$, $\frac{25}{30} = \underline{\quad}$

$\underline{\quad} \square \underline{\quad}$

(2) $\frac{15}{27} = \underline{\quad}$, $\frac{48}{54} = \underline{\quad}$

$\underline{\quad} \square \underline{\quad}$

(4) $\frac{15}{27} = \underline{\quad}$, $\frac{14}{63} = \underline{\quad}$

$\underline{\quad} \square \underline{\quad}$

2 Find the LCM.

2 points per question

(1) (3, 6) → $\square$

(2) (8, 12) → $\square$

(3) (9, 15) → $\square$

(4) (2, 7) → $\square$

3 Add.

5 points per question

(1) $\frac{1}{2} + \frac{3}{5} =$

(3) $1\frac{1}{2} + 3\frac{3}{4} + \frac{1}{8} =$

(2) $\frac{1}{2} + \frac{3}{5} + \frac{1}{6} =$

(4) $1\frac{4}{5} + 3\frac{3}{8} + 4\frac{1}{4} =$

4 Subtract.

5 points per question

(1) $\frac{7}{12}-\frac{1}{3}=$

(3) $4\frac{8}{13}-2\frac{1}{2}=$

(2) $\frac{2}{3}-\frac{1}{4}=$

(4) $3\frac{1}{4}-1\frac{3}{5}=$

5 Calculate.

5 points per question

(1) $\frac{7}{12}-\frac{1}{4}+\frac{1}{3}=$

(2) $\frac{11}{12}+9\frac{1}{6}-\frac{7}{18}=$

6 Multiply.

5 points per question

(1) $\frac{7}{8}\times\frac{3}{4}=$

(3) $5\frac{5}{6}\times2\frac{2}{5}=$

(2) $\frac{10}{11}\times\frac{22}{25}=$

(4) $12\times1\frac{1}{8}\times1\frac{2}{3}=$

7 Divide.

5 points per question

(1) $\frac{3}{8}\div\frac{1}{4}=$

(2) $2\frac{1}{3}\div1\frac{5}{9}=$

You're almost at the finish line!

Review

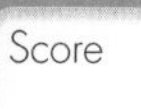

Date / /

Name

Score /100

■ The Answer Key is on page 96.

1 **Rewrite each decimal as a fraction or a mixed number.** 4 points per question

(1) $0.18 =$

(2) $0.125 =$

(3) $8.75 =$

(4) $10.55 =$

2 **Rewrite each fraction or mixed number as a decimal.** 4 points per question

(1) $\frac{4}{5} =$

(2) $\frac{3}{50} =$

(3) $1\frac{2}{5} =$

(4) $3\frac{7}{8} =$

3 **Rewrite each decimal as a fraction and a percent.** 3 points per question

(1) $0.35 = \frac{\quad}{\quad} =$

(2) $0.4805 = \frac{\quad}{\quad} =$

4 **Rewrite each percent as a decimal and a fraction.** 3 points per question

(1) $7\% = \quad = \frac{\quad}{\quad}$

(2) $167.03\% = \quad = \frac{\quad}{\quad}$

5 **Rewrite each fraction as a decimal and a percent.** 3 points per question

(1) $\frac{2}{5} = \quad =$

(2) $2\frac{3}{4} = \quad =$

6 **Calculate.** 5 points per question

(1) $2.75-\frac{2}{3}=$

(2) $2\frac{9}{10}-2.3+1\frac{13}{20}=$

7 **Calculate.** 5 points per question

(1) $\frac{5}{8}\div\frac{8}{9}\times0.8=$

(2) $1.1\div1\frac{13}{15}\times2\frac{2}{3}=$

8 **Calculate.** 5 points per question

(1) $5^3=$

(2) $\left(2\frac{1}{2}\right)^2=$

9 **Calculate.** 5 points per question

(1) $2^4\times\frac{1}{4^2}=$

(2) $\left(1\frac{1}{4}\right)^2\div2\frac{1}{2}=$

10 **Calculate. Write the intermediate steps.** 5 points per question

(1) $\left[\left(1\frac{2}{3}+1.75\right)+3\right]-\frac{1}{2^2}=$

(2) $\frac{1}{3}\div\left[\left(\frac{19}{20}-0.65\right)^2\div0.54\right]=$

Congratulations on completing *Pre-Algebra Workbook I*!

Answer Key Grades 6-8 Pre-Algebra Workbook I

1 Fractions Review pp 2, 3

1 (1) $1\frac{1}{4}$ (2) $1\frac{2}{5}$ (3) 1 (4) $2\frac{1}{3}$ (5) $1\frac{3}{4}$ (6) 2 (7) $2\frac{1}{2}$ (8) $2\frac{1}{4}$ (9) $3\frac{1}{3}$ (10) $1\frac{4}{11}$ (11) 1 (12) $1\frac{6}{7}$ (13) 2 (14) 4 (15) $2\frac{3}{4}$ (16) $4\frac{1}{3}$ (17) $2\frac{1}{7}$ (18) 3 (19) $1\frac{8}{9}$ (20) $2\frac{4}{7}$

2 (1) $\frac{4}{4}$ (2) $\frac{\boxed{7}}{7}$ (3) $\frac{\boxed{10}}{5}$ (4) $\frac{\boxed{14}}{7}$ (5) $\frac{\boxed{9}}{3}$ (6) $\frac{\boxed{12}}{4}$ (7) $\frac{\boxed{9}}{9}$ (8) $\frac{\boxed{16}}{8}$

3 (1) $\frac{3}{2}$ (2) $\frac{9}{5}$ (3) $\frac{14}{5}$ (4) $\frac{17}{6}$ (5) $\frac{11}{7}$ (6) $\frac{7}{3}$ (7) $\frac{14}{3}$ (8) $\frac{20}{13}$ (9) $\frac{38}{11}$ (10) $\frac{17}{8}$ (11) $\frac{15}{11}$ (12) $\frac{23}{4}$

2 Reduction Review pp 4, 5

1 (1) $\frac{1}{2}$ (2) $\frac{1}{3}$ (3) $\frac{4}{5}$ (4) $\frac{7}{10}$ (5) $\frac{7}{10}$

2 (1) $\frac{1}{3}$ (2) $\frac{6}{7}$ (3) $\frac{3}{5}$ (4) $\frac{4}{7}$ (5) $\frac{10}{11}$

3 (1) $\frac{1}{2}$ (2) $\frac{3}{4}$ (3) $\frac{5}{7}$ (4) $\frac{5}{7}$ (5) $\frac{5}{6}$ (6) $\frac{1}{3}$ (7) $\frac{5}{11}$ (8) $\frac{1}{4}$

4 (1) $\frac{3}{5}$ (2) $\frac{5}{6}$ (3) $\frac{1}{7}$ (4) $\frac{1}{2}$ (5) $\frac{1}{3}$ (6) $\frac{1}{2}$ (7) $\frac{3}{7}$ (8) $\frac{7}{15}$ (9) $\frac{1}{3}$ (10) $\frac{1}{3}$ (11) $\frac{2}{5}$ (12) $\frac{3}{8}$

3 Greatest Common Factor pp 6, 7

1 (1) 8 (2) 5 (3) 4 (4) 6 (5) 12 (6) 6

2 (1) 4 (2) 12

3 (1) 6 (2) 5 (3) 14 (4) 9

4 (1) 12, $\frac{1}{2}$ (2) 4, $\frac{4}{5}$ (3) 4, $\frac{4}{7}$ (4) 7, $\frac{2}{5}$ (5) 15, $\frac{1}{2}$ (6) 10, $\frac{1}{4}$ (7) 18, $\frac{1}{2}$ (8) 3, $\frac{3}{11}$ (9) 8, $\frac{1}{7}$ (10) 9, $\frac{1}{6}$ (11) 12, $\frac{3}{4}$ (12) 10, $\frac{2}{7}$ (13) 11, $\frac{2}{5}$ (14) 5, $\frac{7}{12}$ (15) 23, $\frac{1}{2}$ (16) 5, $\frac{5}{13}$ (17) 7, $\frac{2}{3}$ (18) 9, $\frac{2}{5}$ (19) 10, $\frac{2}{9}$ (20) 29, $\frac{1}{3}$

4 Comparing Fractions pp 8, 9

1 (1) $\frac{4}{5}, \frac{3}{5}$; $\frac{4}{5}>\frac{3}{5}$ (2) $\frac{1}{10}, \frac{3}{10}$; $\frac{1}{10}<\frac{3}{10}$ (3) $\frac{11}{15}, \frac{7}{15}$; $\frac{11}{15}>\frac{7}{15}$ (4) $\frac{5}{9}, \frac{8}{9}$; $\frac{5}{9}<\frac{8}{9}$ (5) $\frac{3}{7}, \frac{3}{7}$; $\frac{3}{7}=\frac{3}{7}$ (6) $\frac{2}{13}, \frac{8}{13}$; $\frac{2}{13}<\frac{8}{13}$ (7) $\frac{1}{6}, \frac{5}{6}$; $\frac{1}{6}<\frac{5}{6}$ (8) $\frac{4}{7}, \frac{4}{7}$; $\frac{4}{7}=\frac{4}{7}$ (9) $\frac{5}{9}, \frac{2}{9}$; $\frac{5}{9}>\frac{2}{9}$ (10) $\frac{7}{12}, \frac{11}{12}$; $\frac{7}{12}<\frac{11}{12}$

2 (1) $\frac{2}{3}>\frac{1}{3}$ (2) $\frac{2}{5}=\frac{2}{5}$ (3) $\frac{1}{4}<\frac{3}{4}$ (4) $\frac{6}{13}>\frac{2}{13}$ (5) $\frac{5}{8}>\frac{3}{8}$ (6) $\frac{4}{7}=\frac{4}{7}$

3 (1) $\frac{15}{16}>\frac{12}{16}$ **Ans.** Amanda (2) $\frac{12}{36}, \frac{18}{27} \rightarrow \frac{1}{3}<\frac{2}{3}$ **Ans.** Joey

5 Least Common Multiple pp 10, 11

1 16, 20, 24, 28, 32, 36

2 24, 30, 36, 42, 48, 54

3 21, 28, 35, 42, 49, 56, 63

4 18, 27, 36, 45, 54, 63, 72, 81

5 12, 24, 36

6 18, 36, 54

7 28

8 (1) 12
(2) 18
(3) 28
(4) 36

9 (1) 24 (2) 20 (3) 18 (4) 12 (5) 24 (6) 12 (7) 24 (8) 35 (9) 36 (10) 30 (11) 56 (12) 45 (13) 33 (14) 14 (15) 63 (16) 60

6 Comparing Fractions

pp 12, 13

1 (1) 12 $\frac{4}{12}<\frac{6}{12}$
(2) 20 $\frac{15}{20}>\frac{14}{20}$
(3) 24 $\frac{9}{24}<\frac{10}{24}$
(4) 35 $\frac{14}{35}<\frac{30}{35}$
(5) 12 $\frac{2}{12}<\frac{5}{12}$
(6) 33 $\frac{22}{33}<\frac{24}{33}$
(7) 60 $\frac{48}{60}>\frac{35}{60}$
(8) 10 $\frac{8}{10}<\frac{9}{10}$

2 (1) $\frac{3}{12}<\frac{10}{12}$ (2) $\frac{27}{45}<\frac{35}{45}$ (3) $\frac{40}{45}>\frac{36}{45}$ (4) $\frac{15}{18}>\frac{14}{18}$ (5) $\frac{21}{28}>\frac{12}{28}$ (6) $\frac{13}{26}>\frac{6}{26}$

3 (1) $\frac{5}{7}$, $\frac{10}{11}$ $\frac{55}{77}<\frac{70}{77}$
Ans. Leah

(2) $\frac{3}{4}$, $\frac{8}{9}$ $\frac{27}{36}<\frac{32}{36}$
Ans. Mrs.Castamore's dance class

7 Addition of Fractions

pp 14, 15

1 (1) $\frac{4}{5}$ (2) 1 (3) $1\frac{2}{7}$ (4) $1\frac{4}{9}$ (5) $1\frac{2}{3}$ (6) $1\frac{1}{2}$ (7) $\frac{8}{9}$ (8) $1\frac{2}{11}$ (9) 1 (10) $1\frac{1}{3}$ (11) $1\frac{4}{5}$ (12) $1\frac{1}{2}$

2 (1) 18 (2) 15 (3) 36 (4) 30 (5) 16 (6) 30 (7) 40 (8) 42 (9) 30 (10) 36

3 (1) 12, $\frac{7}{12}$ (2) 24, $\frac{19}{24}$ (3) 56, $\frac{45}{56}$ (4) 36, $\frac{29}{36}$ (5) 24, $\frac{23}{24}$ (6) 45, $\frac{41}{45}$

8 Addition of Fractions

pp 16, 17

1 (1) $\frac{17}{21}$ (2) $\frac{23}{30}$ (3) $1\frac{3}{56}$ (4) $1\frac{23}{42}$ (5) $1\frac{18}{35}$ (6) $\frac{11}{12}$ (7) $\frac{11}{18}$ (8) $1\frac{7}{40}$ (9) $\frac{7}{12}$ (10) $1\frac{13}{24}$ (11) $1\frac{1}{28}$ (12) $1\frac{3}{10}$

2 (1) $1\frac{2}{15}$ (2) $\frac{5}{6}$ (3) $1\frac{11}{21}$ (4) $1\frac{1}{8}$ (5) $1\frac{1}{24}$ (6) $\frac{19}{24}$ (7) $1\frac{5}{12}$ (8) $\frac{9}{20}$ (9) $1\frac{1}{2}$ (10) $\frac{29}{56}$ (11) $\frac{31}{42}$ (12) $\frac{13}{15}$ (13) $\frac{7}{8}$

9 Addition of Fractions

pp 18, 19

1 (1) $\frac{\boxed{2}}{12}+\frac{\boxed{4}}{12}+\frac{\boxed{6}}{12}=\frac{\boxed{12}}{12}=1$
(2) $1\frac{13}{30}$ (3) $1\frac{11}{18}$ (4) $2\frac{1}{5}$ (5) $1\frac{17}{18}$ (6) $1\frac{7}{60}$ (7) $1\frac{1}{36}$ (8) $1\frac{17}{40}$ (9) $1\frac{23}{44}$ (10) $1\frac{1}{5}$

2 (1) $1\frac{14}{15}$
(2) $1\frac{\boxed{4}}{8}+3\frac{\boxed{6}}{8}+\frac{1}{8}=5\frac{3}{8}$
(3) $3\frac{2}{3}$ (4) $3\frac{59}{70}$ (5) $9\frac{17}{40}$ (6) $4\frac{11}{24}$ (7) $5\frac{1}{12}$ (8) $4\frac{1}{12}$ (9) $4\frac{5}{36}$ (10) $3\frac{5}{8}$

10 Subtraction of Fractions

pp 20, 21

1 (1) $\frac{2}{7}$ (2) $\frac{1}{2}$ (3) $\frac{2}{3}$ (4) $\frac{5}{13}$ (5) $\frac{5}{11}$ (6) $\frac{1}{5}$ (7) $\frac{1}{6}$ (8) $\frac{1}{3}$ (9) $\frac{3}{10}$ (10) $\frac{5}{12}$

2 (1) $\frac{5}{12}$ (2) $\frac{1}{4}$ (3) $\frac{3}{22}$
(4) $2\frac{\boxed{14}}{22}-1\frac{\boxed{11}}{22}=1\frac{3}{22}$
(5) $2\frac{8}{21}$ (6) $2\frac{1}{12}$ (7) $1\frac{3}{10}$ (8) $2\frac{1}{3}$ (9) $2\frac{5}{18}$ (10) $1\frac{8}{15}$

11 Subtraction of Fractions
pp 22, 23

1
(1) $3\frac{\boxed{5}}{20}-1\frac{\boxed{12}}{20}=2\frac{\boxed{25}}{20}-1\frac{\boxed{12}}{20}=1\frac{13}{20}$
(2) $2\frac{19}{21}$
(3) $3\frac{11}{14}$
(4) $2\frac{7}{10}$
(5) $2\frac{13}{22}$
(6) $1\frac{23}{26}$
(7) $4\frac{13}{16}$
(8) $\frac{27}{44}$
(9) $1\frac{5}{14}$
(10) $\frac{7}{20}$

2
(1) $\frac{\boxed{11}}{12}-\frac{\boxed{3}}{12}-\frac{\boxed{4}}{12}=\frac{4}{12}=\frac{1}{3}$
(2) $\frac{1}{7}$
(3) $2\frac{7}{24}$
(4) $3\frac{7}{24}$
(5) $5\frac{3}{4}$
(6) $2\frac{7}{18}$
(7) $2\frac{1}{24}$
(8) $3\frac{5}{24}$
(9) $2\frac{5}{36}$
(10) $\frac{7}{8}$

12 Addition & Subtraction of Fractions
pp 24, 25

1
(1) $\frac{\boxed{8}}{18}+\frac{\boxed{9}}{18}-\frac{\boxed{12}}{18}=\frac{5}{18}$
(2) $\frac{2}{3}$
(3) $4\frac{1}{18}$
(4) $2\frac{1}{2}$
(5) $9\frac{8}{45}$
(6) $5\frac{3}{5}$
(7) $7\frac{1}{3}$
(8) $1\frac{2}{3}$
(9) $11\frac{5}{18}$
(10) $10\frac{13}{21}$

2
(1) $\frac{7}{8}$
(2) $2\frac{7}{12}$
(3) $7\frac{1}{3}$
(4) $3\frac{17}{18}$
(5) $8\frac{31}{36}$
(6) $7\frac{3}{28}$
(7) $1\frac{7}{12}$
(8) $1\frac{9}{10}$
(9) $1\frac{5}{24}$
(10) $2\frac{19}{72}$

13 Word Problems with Fractions
pp 26, 27

1
(1) $\frac{4}{10}+\frac{1}{10}=\frac{5}{10}=\frac{1}{2}$ **Ans.** $\frac{1}{2}$ liter
(2) $\frac{1}{9}+\frac{1}{3}=\frac{1}{9}+\frac{3}{9}=\frac{4}{9}$ **Ans.** $\frac{4}{9}$ meter
(3) $\frac{3}{6}+\frac{4}{6}+\frac{1}{6}=\frac{8}{6}=1\frac{2}{6}=1\frac{1}{3}$ **Ans.** $1\frac{1}{3}$ kilometers
(4) $1\frac{1}{3}+\frac{1}{2}+2\frac{5}{6}=1\frac{2}{6}+\frac{3}{6}+2\frac{5}{6}=3\frac{10}{6}=4\frac{4}{6}=4\frac{2}{3}$
Ans. $4\frac{2}{3}$ cups

2
(1) $3\frac{3}{4}-1\frac{2}{5}=3\frac{\boxed{15}}{20}-1\frac{\boxed{8}}{20}=2\frac{7}{20}$ **Ans.** $2\frac{7}{20}$ pounds
(2) $2\frac{8}{9}-1\frac{1}{3}=2\frac{8}{9}-1\frac{3}{9}=1\frac{5}{9}$ **Ans.** $1\frac{5}{9}$ cups
(3) $8\frac{4}{15}-5\frac{7}{10}-1\frac{1}{5}=1\frac{11}{30}$ **Ans.** $1\frac{11}{30}$ ounces
(4) $15\frac{3}{4}-8\frac{7}{12}-2\frac{1}{3}=4\frac{5}{6}$ **Ans.** $4\frac{5}{6}$ gallons

3
(1) $1\frac{7}{15}-\frac{4}{5}+\frac{1}{3}=1$ **Ans.** 1 pound
(2) $4\frac{1}{6}-\frac{5}{12}+\frac{1}{2}=4\frac{1}{4}$ **Ans.** $4\frac{1}{4}$ pages

14 Multiplication of Fractions
pp 28, 29

1
(1) $\frac{\boxed{1}}{12}$
(2) $\frac{18}{35}$
(3) $\frac{16}{63}$
(4) $\frac{8}{25}$
(5) $\frac{21}{32}$
(6) $\frac{20}{63}$
(7) $\frac{14}{27}$
(8) $\frac{11}{26}$
(9) $\frac{11}{48}$
(10) $\frac{25}{84}$

2
(1) $\frac{2}{5}$
(2) $\frac{\cancel{10}^{\boxed{2}}}{\cancel{11}_{\boxed{1}}}\times\frac{\cancel{22}^{\boxed{2}}}{\cancel{15}_{\boxed{5}}}=\frac{\boxed{4}}{\boxed{5}}$
(3) $\frac{4}{27}$
(4) $\frac{5}{44}$
(5) $\frac{9}{50}$
(6) $\frac{5}{6}$
(7) $\frac{2}{3}$
(8) $\frac{21}{23}$
(9) $\frac{3}{44}$
(10) $\frac{5}{8}$

15 Multiplication of Fractions
pp 30, 31

1
(1) $\frac{5}{\cancel{3}_{\boxed{1}}}\times\frac{\cancel{9}^{\boxed{3}}}{4}=\frac{15}{4}=3\frac{3}{4}$
(2) $\frac{9}{\cancel{4}_{1}}\times\frac{\cancel{28}^{7}}{13}=\frac{63}{13}=4\frac{11}{13}$
(3) 18
(4) $8\frac{6}{7}$
(5) $5\frac{1}{4}$
(6) 24
(7) 14
(8) $10\frac{1}{2}$
(9) 35
(10) 9

2
(1) $\frac{1}{6}$
(2) $\frac{2}{15}$
(3) $\frac{4}{5}$
(4) $19\frac{1}{2}$
(5) $27\frac{1}{2}$
(6) $4\frac{1}{4}$
(7) $13\frac{1}{2}$
(8) $\frac{4}{5}$
(9) $22\frac{1}{2}$
(10) $\frac{13}{30}$

16 Division of Fractions
pp 32, 33

1 (1) $\frac{8}{9}\times\frac{\boxed{2}}{1}=\frac{16}{9}=1\frac{7}{9}$

(2) $\frac{4}{5}\times\frac{\boxed{7}}{\boxed{3}}=\frac{28}{15}=1\frac{13}{15}$

(3) $\frac{9}{20}$

(4) $1\frac{17}{18}$

(5) $\frac{16}{35}$

(6) $\frac{18}{35}$

(7) $1\frac{5}{16}$

(8) $\frac{27}{35}$

(9) $3\frac{3}{14}$

(10) $2\frac{10}{13}$

2 (1) $1\frac{1}{7}$

(2) $\frac{2}{3}$

(3) $1\frac{1}{9}$

(4) $\frac{5}{7}$

(5) $\frac{3}{7}$

(6) $1\frac{1}{2}$

(7) $1\frac{1}{5}$

(8) $1\frac{1}{14}$

(9) $1\frac{1}{20}$

(10) $\frac{44}{45}$

17 Division of Fractions
pp 34, 35

1 (1) $\frac{1}{7}\times\frac{1}{\boxed{4}}=\frac{1}{28}$

(2) $\frac{6}{7}\div\frac{\boxed{7}}{1}=\frac{6}{7}\times\frac{1}{7}=\frac{6}{49}$

(3) $\frac{3}{40}$

(4) $\frac{\boxed{5}}{1}\div\frac{1}{4}=\frac{5}{1}\times\frac{4}{1}=20$

(5) $4\frac{4}{7}$

2 (1) $\frac{5}{6}$

(2) $\frac{8}{33}$

(3) 2

(4) $1\frac{1}{3}$

(5) $1\frac{1}{2}$

3 (1) $\frac{4}{27}$

(2) $\frac{4}{5}\div\frac{\boxed{5}}{3}\div\frac{3}{\boxed{2}}=\frac{4}{5}\times\frac{3}{5}\times\frac{2}{3}=\frac{8}{25}$

(3) $\frac{1}{8}$

(4) $\frac{2}{45}$

(5) $1\frac{2}{7}$

(6) $2\frac{5}{8}$

4 (1) 1

(2) $\frac{1}{3}$

(3) $\frac{8}{15}$

(4) 2

18 Word Problems with Fractions
pp 36, 37

1 (1) $1\frac{1}{4}\times 6=7\frac{1}{2}$ **Ans.** $7\frac{1}{2}$ miles

(2) $2\frac{4}{9}\times 12=29\frac{1}{3}$ **Ans.** $29\frac{1}{3}$ pounds

(3) $4\frac{1}{4}\div\frac{2}{3}\times 6=38\frac{1}{4}$ **Ans.** $38\frac{1}{4}$ feet

(4) $3\times\frac{1}{2}\times\frac{1}{2}=\frac{3}{4}$ **Ans.** $\frac{3}{4}$ cup

2 (1) $\frac{5}{7}\times\frac{1}{2}=\frac{5}{14}$ **Ans.** $\frac{5}{14}$ yard

(2) $\frac{6}{7}\div 3=\frac{6}{7}\times\frac{1}{3}=\frac{2}{7}$ **Ans.** $\frac{2}{7}$ pound

(3) $12\div\frac{3}{5}=20$ **Ans.** 20 liters

(4) $1\frac{14}{15}\div 2=\frac{29}{30}$ **Ans.** $\frac{29}{30}$ liter

19 Fraction Review
pp 38, 39

1 (1) $1\frac{18}{35}$

(2) $3\frac{59}{70}$

(3) $2\frac{19}{21}$

(4) $1\frac{1}{7}$

(5) $4\frac{1}{2}$

(6) $3\frac{8}{45}$

2 (1) $8\frac{6}{7}$

(2) $12\frac{1}{2}$

(3) $\frac{18}{23}$

(4) $\frac{11}{36}$

(5) $18\frac{2}{3}$

(6) $\frac{5}{32}$

3 (1) $1\frac{3}{4}+1\frac{1}{3}+1\frac{1}{6}=4\frac{1}{4}$ **Ans.** $4\frac{1}{4}$ pounds

(2) $5\frac{1}{9}+1\frac{1}{3}-1\frac{5}{18}=5\frac{1}{6}$ **Ans.** $5\frac{1}{6}$ cups

4 (1) $4\frac{1}{4}\div\frac{7}{8}\times 4\frac{1}{2}=21\frac{6}{7}$ **Ans.** $21\frac{6}{7}$ miles

(2) $7\frac{1}{2}\div 4\frac{1}{6}\times 8\frac{3}{4}=15\frac{3}{4}$ **Ans.** $15\frac{3}{4}$ yards

20 Place Value Review
pp 40, 41

1 (1) 4, 6 (2) 2, 4 (3) 8, 2 (4) 9, 3 (5) 7, 4

2 (1) (From the left) 0.1, 0.4, 0.7

(2) 0.07, 0.25, 0.61, 0.85, 0.99

(3) 0.19, 0.36, 0.56, 1.01

3 (1) 6.7 (2) 8.5 (3) 0.2

4 (1) 8.92 (2) 4.45 (3) 0.11

5 (1) 0.157 (2) 0.112 (3) 23.783

6 (1) 0.015 (2) 0.015 (3) 0.014 (4) 0.016

21 Decimals as Fractions
pp 42, 43

1 (1) $\frac{\boxed{5}}{\boxed{10}}=\frac{\boxed{1}}{\boxed{2}}$

(2) $\frac{\boxed{8}}{\boxed{10}}=\frac{\boxed{4}}{\boxed{5}}$

(3) $\frac{\boxed{25}}{\boxed{100}}=\frac{\boxed{1}}{\boxed{4}}$

(4) $\frac{\boxed{35}}{\boxed{100}}=\frac{\boxed{7}}{\boxed{20}}$

(5) $\frac{\boxed{5}}{\boxed{100}}=\frac{\boxed{1}}{\boxed{20}}$

(6) $\frac{\boxed{18}}{\boxed{100}}=\frac{\boxed{9}}{\boxed{50}}$

(7) $\frac{\boxed{8}}{\boxed{100}}=\frac{\boxed{2}}{\boxed{25}}$

(8) $\frac{\boxed{5}}{\boxed{1000}}=\frac{\boxed{1}}{\boxed{200}}$

(9) $\frac{\boxed{825}}{\boxed{1000}}=\frac{\boxed{33}}{\boxed{40}}$

(10) $\frac{\boxed{404}}{\boxed{1000}}=\frac{\boxed{101}}{\boxed{250}}$

2 (1) $3\frac{1}{2}$ (2) $5\frac{3}{4}$ (3) $2\frac{3}{5}$ (4) $2\frac{17}{25}$ (5) $10\frac{11}{20}$ (6) $9\frac{1}{200}$

3 (1) $\frac{5}{2}$ (2) $\frac{17}{5}$ (3) $\frac{28}{25}$ (4) $\frac{52}{25}$ (5) $\frac{13}{8}$ (6) $\frac{251}{125}$

22 Fractions as Decimals

pp 44, 45

1 (1) 0.25

$$\begin{array}{r} 0.25 \\ 4\overline{)1.0} \\ 8 \\ \hline 20 \\ 20 \\ \hline 0 \end{array}$$

(2) 0.75 (3) 0.8 (4) 1.4 (5) 0.06 (6) 0.375 (7) 8.25 (8) 7.75

2 (1) 2.25 (2) 7.8 (3) 5.375 (4) 4.12 (5) 3.62 (6) 10.2 (7) 5.48 (8) 2.6 (9) 7.875 (10) 2.625

23 Percents

pp 46, 47

1 (1) $\boxed{45}$, 45% (2) 35, 35% (3) 40, 40% (4) 4, 4% (5) 4, 0.4% (6) 123, 12.3% (7) 971, 97.1% (8) 979, 97.9% (9) 4805, 48.05% (10) 3, 0.03%

2 (1) 100% (2) 145% (3) 104% (4) 104.5% (5) 1047% (6) 100.5% (7) 101.2% (8) 110.3% (9) 200% (10) 202%

24 Percents

pp 48, 49

1 (1) $\frac{\boxed{45}}{\boxed{100}}=0.45$ (2) $\frac{\boxed{55}}{\boxed{100}}=0.55$ (3) $\frac{\boxed{70}}{\boxed{100}}=0.7$ (4) $\frac{\boxed{87}}{\boxed{100}}=0.87$ (5) $\frac{\boxed{6}}{\boxed{100}}=0.06$ (6) $\frac{\boxed{264}}{\boxed{1000}}=0.264$ (7) $\frac{\boxed{149}}{\boxed{1000}}=0.149$ (8) $\frac{\boxed{8}}{\boxed{1000}}=0.008$ (9) $\frac{\boxed{6702}}{\boxed{10000}}=0.6702$ (10) $\frac{\boxed{9}}{\boxed{10000}}=0.0009$

2 (1) 1 (2) 1.45 (3) 1.08 (4) 1.17 (5) 1.1 (6) 10 (7) 11 (8) 11.1 (9) 11.11 (10) 10.11

25 Percents

pp 50, 51

1 (1) $0.25=\frac{\boxed{25}}{\boxed{100}}=\frac{\boxed{1}}{\boxed{4}}$ (2) $0.75=\frac{75}{100}=\frac{3}{4}$ (3) $0.6=\frac{60}{100}=\frac{3}{5}$ (4) $0.15=\frac{15}{100}=\frac{3}{20}$ (5) $0.26=\frac{26}{100}=\frac{13}{50}$ (6) $1.5=\frac{15}{10}=\frac{3}{2}=1\frac{1}{2}$ (7) $1.25=\frac{125}{100}=\frac{5}{4}=1\frac{1}{4}$ (8) $1.05=\frac{105}{100}=\frac{21}{20}=1\frac{1}{20}$ (9) $0.105=\frac{105}{1000}=\frac{21}{200}$ (10) $1.005=\frac{1005}{1000}=\frac{201}{200}=1\frac{1}{200}$

2 (1) 0.5=50% (2) 0.25=25% (3) 0.4=40% (4) 0.35=35% (5) 1.8=180% (6) 1.75=175%

3 (1) 0.82=82% **Ans.** 82%
(2) 1.47=147% **Ans.** 147%

4 (1) 8,250%=82.5 **Ans.** 82.5
(2) 27%=0.27 **Ans.** 0.27

5 (1) $\frac{5}{8}=0.625=62.5\%$ **Ans.** 62.5%
(2) $\frac{1}{5}=0.2=20\%$ **Ans.** 20%

26 Decimals and Fractions

pp 52, 53

1

(1) $\frac{1}{6}+\frac{1}{2}=\frac{4}{6}=\frac{2}{3}$

(2) $\frac{13}{20}$

(3) $1\frac{9}{20}$

(4) $2\frac{1}{3}$

(5) $1\frac{29}{40}$

(6) $\frac{5}{12}$

(7) $\frac{3}{10}$

(8) $\frac{7}{30}$

(9) $2\frac{1}{12}$

(10) $3\frac{7}{20}$

2

(1) $\frac{9}{10}$

(2) $1\frac{9}{10}$

(3) $\frac{39}{40}$

(4) $5\frac{3}{20}$

(5) $2\frac{17}{20}$

(6) $1\frac{19}{20}$

(7) $\frac{4}{5}$

(8) $\frac{23}{24}$

(9) $1\frac{19}{24}$

(10) $\frac{13}{36}$

27 Decimals and Fractions

pp 54, 55

1

(1) $\frac{1}{12}$

(2) $\frac{5}{8}$

(3) $\frac{1}{30}$

(4) $2\frac{2}{5}$

(5) $\frac{1}{15}$

(6) $1\frac{4}{5}$

(7) $\frac{11}{18}$

(8) $1\frac{19}{20}$

(9) $8\frac{1}{3}$

(10) $\frac{3}{10}$

2

(1) $\frac{1}{2}$

(2) $\frac{9}{16}$

(3) $\frac{2}{3}$

(4) $\frac{20}{21}$

(5) $\frac{5}{8}$

(6) $5\frac{5}{8}$

(7) $2\frac{14}{25}$

(8) $2\frac{33}{50}$

(9) $11\frac{1}{5}$

(10) $\frac{17}{25}$

28 Word Problems with Decimals and Fractions

pp 56, 57

1

(1) $\frac{9}{10}-0.15+\frac{3}{100}=\frac{39}{50}$ **Ans.** $\frac{39}{50}$ liter

(2) $6\frac{9}{10}-2.15+\frac{1}{6}=4\frac{11}{12}$ **Ans.** $4\frac{11}{12}$ feet

2

(1) $1\frac{8}{15}\times4.5=6\frac{9}{10}$ **Ans.** $6\frac{9}{10}$ yards

(2) $2.15\times3\frac{1}{3}=7\frac{1}{6}$ **Ans.** $7\frac{1}{6}$ meters

3

(1) $3\frac{1}{3}\times2.5\div\frac{5}{6}=10$ **Ans.** 10 pieces

(2) $1.2\times6.75\div\frac{9}{10}=9$ **Ans.** 9 pieces

4

(1) $0.6\times\frac{1}{9}+0.1=\frac{1}{6}$ **Ans.** $\frac{1}{6}$ ton

(2) $1.5\times\frac{9}{10}-0.25=1\frac{1}{10}$ **Ans.** $1\frac{1}{10}$ liters

29 Exponents

pp 58, 59

1

(1) 4

(2) 8

(3) 16

(4) 9

(5) 27

(6) 1

(7) 243

(8) 256

(9) 1

(10) 343

2

(1) $\frac{1}{4}$

(2) $\frac{1}{8}$

(3) $\frac{1}{16}$

(4) $\frac{1}{16}$

(5) $\frac{3}{125}$

(6) 1

3

(1) $\frac{3}{4}\times\frac{3}{4}\times\frac{3}{4}=\frac{27}{64}$ **Ans.** $\frac{27}{64}$ in^3

(2) $\frac{4}{5}\times\frac{4}{5}\times\frac{4}{5}=\frac{64}{125}$ **Ans.** $\frac{64}{125}$ ft^3

30 Exponents

pp 60, 61

1

(1) $1\frac{7}{9}$

(2) $2\frac{10}{27}$

(3) $3\frac{3}{8}$

(4) $5\frac{1}{16}$

(5) $6\frac{1}{4}$

(6) $2\frac{7}{9}$

(7) $5\frac{1}{16}$

(8) $4\frac{17}{27}$

(9) 1

(10) $14\frac{1}{16}$

2

(1) 8

(2) $\frac{3}{8}$

(3) 4

(4) $\frac{5}{8}$

(5) 12

(6) 144

3

(1) $\frac{3}{4}$

(2) $121\frac{1}{2}$

(3) 2500

(4) 256

(5) 225

(6) $\frac{25}{216}$

31 Order of Operations

pp 62, 63

1

(1) $7-\boxed{5}=2$

(2) 4

(3) 9

(4) 9

(5) 7

(6) 56

(7) 4

(8) 6

(9) 4

(10) 6

2
(1) $\frac{5}{7}-\frac{\boxed{4}}{7}=\frac{1}{7}$
(2) $\frac{1}{2}+\left(\frac{\boxed{20}}{4}-\frac{1}{4}\right)=5\frac{1}{4}$
(3) $\frac{2}{3}$
(4) $\frac{2}{5}$
(5) 15
(6) $2\frac{1}{6}$
(7) 14
(8) $\frac{3}{4}$
(9) 3
(10) $5\frac{1}{2}$

32 Order of Operations

pp 64, 65

1
(1) $[4+\boxed{7}]-9=\boxed{11}-9=2$
(2) 7
(3) 9
(4) 12
(5) $1\frac{1}{6}$
(6) $[\boxed{3}\div\boxed{1}]+\frac{5}{7}=\boxed{3}+\frac{5}{7}=3\frac{5}{7}$
(7) 3
(8) $3\frac{27}{28}$
(9) 4
(10) $5\frac{6}{7}$

2
(1) $6-\left[\left(\frac{1}{4}+\frac{\boxed{3}}{\boxed{4}}\right)-\frac{3}{4}\right]=5\frac{3}{4}$
(2) $4-\left[\frac{1}{3}+1\frac{\boxed{1}}{\boxed{10}}\right]=2\frac{17}{30}$
(3) $\frac{4}{5}$ (or 0.8)
(4) $6\frac{1}{2}$ (or 6.5)
(5) $7\frac{2}{5}$
(6) 6
(7) 6
(8) 30
(9) 10
(10) $9\frac{1}{15}$

33 Order of Operations

pp 66, 67

1
(1) $3^2-(4+\boxed{5})=3^2-\boxed{9}=\boxed{9}-\boxed{9}=0$
(2) 32
(3) 19
(4) 2
(5) $[2+(\boxed{9}-4)]\times\frac{5}{7}=[2+\boxed{5}]\times\frac{5}{7}=5$
(6) $\frac{6}{7}$
(7) $8\frac{3}{16}$
(8) $3\frac{1}{2}$

2
(1) $16\frac{5}{18}$
(2) $11\frac{25}{49}$
(3) 3
(4) $\frac{2}{3}$
(5) $\frac{11}{81}$
(6) 6
(7) $6\frac{11}{20}$
(8) $51\frac{7}{9}$
(9) $5\frac{9}{64}$
(10) 32

34 Order of Operations

pp 68, 69

1
(1) $8\frac{1}{4}$
(2) $14\frac{7}{12}$
(3) 8
(4) $16\frac{23}{36}$
(5) 10
(6) $9\frac{3}{5}$
(7) 36
(8) 8

2
(1) 18
(2) 10
(3) $13\frac{1}{24}$
(4) 2
(5) $9\frac{1}{8}$
(6) 27

35 Order of Operations pp70,71

1

(1) $6-\boxed{4}=2$
(2) 10
(3) 11
(4) $\frac{2}{3}+\frac{\boxed{2}}{\boxed{3}}=\frac{4}{3}=1\frac{1}{3}$
(5) $7\frac{3}{4}$
(6) 2
(7) 10
(8) $9\frac{1}{2}$
(9) $2\frac{2}{9}$
(10) $1\frac{1}{6}$

2

(1) $1\frac{1}{2}$
(2) $1\frac{1}{2}$
(3) $1\frac{7}{10}$
(4) $3\frac{1}{3}$
(5) $9\frac{1}{3}$
(6) $\frac{29}{30}$
(7) $\frac{3}{4}$
(8) $2\frac{1}{7}$
(9) 7
(10) $\frac{1}{2}$

36 Word Problems with Order of Operations pp72,73

1

(1) $\boxed{15}-(\boxed{16}-\boxed{3})=15-13=2$ **Ans.** 2 cupcakes
(2) $(\boxed{8}+\boxed{8})-(\boxed{9}+\boxed{5})=16-14=2$ **Ans.** 2 bracelets

2

(1) $\boxed{1\frac{1}{4}}^2+\boxed{6}^2=37\frac{9}{16}$ **Ans.** $37\frac{9}{16}$ yd²
(2) $\left(\boxed{2}\times\boxed{\frac{1}{2}}\right)+\boxed{4}^2+\boxed{4}^2=33$ **Ans.** 33 yd²

3

(1) $1\frac{5}{8}\times4-1\frac{3}{4}=4\frac{3}{4}$ **Ans.** $4\frac{3}{4}$ cups
(2) $2\frac{1}{4}\times5\div4=2\frac{13}{16}$ **Ans.** $2\frac{13}{16}$ cups

4

(1) $2\times3\frac{5}{6}-1\frac{1}{2}=6\frac{1}{6}$ **Ans.** $6\frac{1}{6}$ kilograms
(2)) $\left(\frac{1}{6}+\frac{1}{4}\right)\times2=\frac{5}{6}$ **Ans.** $\frac{5}{6}$ of his book

37 Order of Operations pp74,75

1

(1) $3+[\boxed{4}\times\boxed{3}]\div2=3+\boxed{12}\div2=3+\boxed{6}=9$
(2) 6
(3) 4
(4) 9
(5) 4
(6) 18
(7) 6
(8) 5
(9) 53
(10) 12

2

(1) 32
(2) 1
(3) 4
(4) 44
(5) 6
(6) 18
(7) 5
(8) 5
(9) 80
(10) 2

38 Order of Operations pp76,77

1

(1) $\left(\frac{\boxed{3}}{\boxed{4}}\times\frac{\boxed{3}}{\boxed{4}}\right)+(\boxed{35}-3)=\frac{\boxed{9}}{\boxed{16}}+\boxed{32}=32\frac{9}{16}$
(2) $4\frac{5}{14}$
(3) $4\frac{16}{25}$
(4) 14
(5) 35
(6) 12
(7) $2\frac{3}{5}$
(8) 2
(9) $\frac{1}{12}$
(10) $7\frac{49}{64}$

2

(1) $\frac{3}{\boxed{4}\times\boxed{4}}+[7\times2]=14\frac{3}{16}$
(2) $33\frac{1}{3}$
(3) 51
(4) 6
(5) 10
(6) $16\frac{7}{18}$
(7) $22\frac{1}{2}$
(8) 41

39 Order of Operations pp78,79

1

(1) 10
(2) $2\frac{3}{5}$
(3) $\frac{1}{45}$
(4) 14
(5) $6\frac{1}{3}$

2

(1) $\frac{14}{15}$
(2) 27
(3) $1\frac{11}{16}$
(4) $4\frac{1}{8}$
(5) $2\frac{1}{2}$
(6) $4\frac{1}{2}$

40 Order of Operations pp80,81

1

(1) $13\frac{1}{3}$
(2) $\frac{9}{25}$
(3) $1\frac{2}{3}$
(4) 9
(5) 1

2

(1) 4
(2) $1\frac{5}{16}$
(3) $1\frac{2}{3}$
(4) $\frac{1}{2}$
(5) 1
(6) 4

41 Order of Operations

pp 82, 83

1 (1) $4\frac{5}{16}$ (2) $5\frac{19}{25}$ (3) $1\frac{2}{3}$ (4) 4 (5) 9

2 (1) 4 (2) $\frac{1}{11}$ (3) $2\frac{5}{8}$ (4) 2 (5) 5 (6) $4\frac{3}{16}$

42 Review

pp 84, 85

1 (1) $\frac{1}{10}, \frac{3}{10}$ $\quad \frac{1}{10} < \frac{3}{10}$

(2) $\frac{5}{9}, \frac{8}{9}$ $\quad \frac{5}{9} < \frac{8}{9}$

(3) $\frac{1}{6}, \frac{5}{6}$ $\quad \frac{1}{6} < \frac{5}{6}$

(4) $\frac{5}{9}, \frac{2}{9}$ $\quad \frac{5}{9} > \frac{2}{9}$

2 (1) 6 (2) 24 (3) 45 (4) 14

3 (1) $1\frac{1}{10}$ (2) $1\frac{4}{15}$ (3) $5\frac{3}{8}$ (4) $9\frac{17}{40}$

4 (1) $\frac{1}{4}$ (2) $\frac{5}{12}$ (3) $2\frac{3}{26}$ (4) $1\frac{13}{20}$

5 (1) $\frac{2}{3}$ (2) $9\frac{25}{36}$

6 (1) $\frac{21}{32}$ (2) $\frac{4}{5}$ (3) 14 (4) $22\frac{1}{2}$

7 (1) $1\frac{1}{2}$ (2) $1\frac{1}{2}$

43 Review

pp 86, 87

1 (1) $\frac{9}{50}$ (2) $\frac{1}{8}$ (3) $8\frac{3}{4}$ (4) $10\frac{11}{20}$

2 (1) 0.8 (2) 0.06 (3) 1.4 (4) 3.875

3 (1) $\frac{35}{100} = 35\%$ (2) $\frac{4805}{10000} = 48.05\%$

4 (1) $0.07 = \frac{7}{100}$ (2) $1.6703 = \frac{16703}{10000}$

5 (1) $0.4 = 40\%$ (2) $2.75 = 275\%$

6 (1) $2\frac{1}{12}$ (2) $2\frac{1}{4}$

7 (1) $\frac{9}{16}$ (2) $1\frac{4}{7}$

8 (1) 125 (2) $6\frac{1}{4}$

9 (1) 1 (2) $\frac{5}{8}$

10 (1) $6\frac{1}{6}$ (2) 2

KUMON MATH WORKBOOKS

Pre-Algebra

Workbook II

Table of Contents

1 Greatest Common Factor Review

Level ☆

Date / /

Name

Score /100

■ The Answer Key is on page 184.

1 **Reduce by using the greatest common factor (GCF).**

2 points per question

(1) $\frac{2}{6} =$

(2) $\frac{2}{4} =$

(3) $\frac{3}{6} =$

(4) $\frac{4}{10} =$

(5) $\frac{4}{6} =$

(6) $\frac{5}{35} =$

(7) $\frac{12}{15} =$

(8) $\frac{15}{40} =$

(9) $\frac{14}{21} =$

(10) $\frac{11}{33} =$

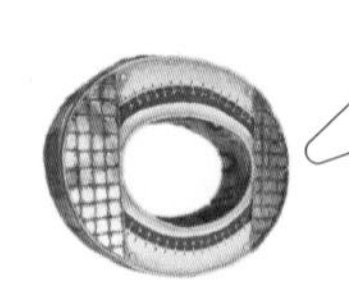

To reduce by the GCF, simplify the fraction by dividing the numerator and denominator by the largest factor they have in common.

2 Reduce by using the GCF.

4 points per question

(1) $\frac{4}{8} =$

(2) $\frac{2}{18} =$

(3) $\frac{6}{18} =$

(4) $\frac{30}{50} =$

(5) $\frac{9}{27} =$

(6) $\frac{10}{60} =$

(7) $\frac{12}{24} =$

(8) $\frac{30}{36} =$

(9) $\frac{18}{63} =$

(10) $\frac{20}{40} =$

(11) $\frac{6}{30} =$

(12) $\frac{21}{35} =$

(13) $\frac{30}{35} =$

(14) $\frac{20}{45} =$

(15) $\frac{16}{56} =$

(16) $\frac{27}{72} =$

(17) $\frac{40}{48} =$

(18) $\frac{13}{26} =$

(19) $\frac{22}{44} =$

(20) $\frac{24}{60} =$

You're off to a great start!

2 Least Common Multiple Review

Level ☆

Date / /

Name

Score /100

■ The Answer Key is on page 184.

1 Write the least common multiple (LCM) of each number pair.

2 points per question

(1) (3, 5) → ☐

(6) (9, 12) → ☐

(2) (5, 8) → ☐

(7) (5, 10) → ☐

(3) (8, 9) → ☐

(8) (3, 12) → ☐

(4) (4, 10) → ☐

(9) (8, 12) → ☐

(5) (6, 8) → ☐

(10) (10, 15) → ☐

The LCM is the lowest multiple that two or more integers have in common.

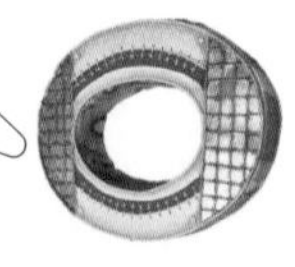

2 Write the LCM of each number pair.

4 points per question

(1) (2, 5) → ☐

(2) (4, 7) → ☐

(3) (4, 6) → ☐

(4) (6, 9) → ☐

(5) (5, 15) → ☐

(6) (5, 7) → ☐

(7) (7, 8) → ☐

(8) (6, 11) → ☐

(9) (12, 18) → ☐

(10) (15, 20) → ☐

(11) (12, 16) → ☐

(12) (8, 10) → ☐

(13) (16, 20) → ☐

(14) (6, 14) → ☐

(15) (10, 14) → ☐

(16) (6, 27) → ☐

(17) (12, 20) → ☐

(18) (40, 60) → ☐

(19) (22, 33) → ☐

(20) (15, 60) → ☐

If you have difficulty with this workbook, please try Kumon's *Pre-Algebra Workbook I*.

3 Addition of Fractions Review

Level ☆

Date / /

Name

Score /100

■ The Answer Key is on page 184.

1 Add.

3 points per question

(1) $\frac{3}{8}+\frac{1}{2}=$

(2) $\frac{7}{10}+\frac{1}{5}=$

(3) $\frac{1}{6}+\frac{1}{2}=$

(4) $\frac{1}{2}+\frac{3}{10}=$

(5) $\frac{1}{3}+\frac{5}{21}=$

(6) $\frac{3}{4}+\frac{1}{5}=$

(7) $\frac{2}{7}+\frac{2}{5}=$

(8) $\frac{3}{4}+\frac{5}{8}=$

(9) $\frac{2}{3}+\frac{4}{5}=$

(10) $\frac{3}{10}+\frac{13}{15}=$

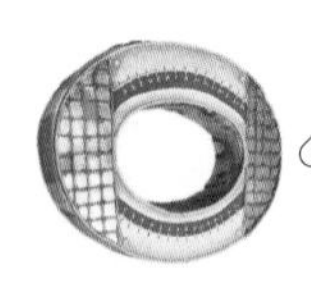

Find the LCM of the denominators to calculate fractions with different denominators.

2 Add.

5 points per question

(1) $6\frac{1}{4}+\frac{3}{8}=$

(2) $2\frac{5}{6}+\frac{1}{8}=$

(3) $\frac{3}{4}+4\frac{5}{8}=$

(4) $2\frac{2}{3}+5\frac{3}{4}=$

(5) $6\frac{7}{10}+1\frac{5}{6}=$

(6) $3\frac{3}{4}+2\frac{11}{12}=$

(7) $\frac{1}{2}+\frac{1}{3}+\frac{1}{4}=$

(8) $\frac{2}{3}+\frac{1}{2}+\frac{4}{5}=$

(9) $1\frac{2}{5}+\frac{1}{2}+2\frac{3}{8}=$

(10) $2\frac{3}{4}+5\frac{7}{10}+1\frac{2}{5}=$

3 Answer each word problem. Write the question as an expression first, and then calculate.

10 points per question

(1) David ate $\frac{2}{3}$ of a pizza, while Grace ate $\frac{1}{4}$ of the pizza. How much of the pizza did they eat altogether?

⟨Ans.⟩

(2) Jay walked $1\frac{4}{5}$ miles on Monday, $2\frac{1}{2}$ miles on Tuesday, and $\frac{1}{6}$ mile on Wednesday. How far did he walk altogether?

Wow! You are good at this!

⟨Ans.⟩

Subtraction of Fractions Review

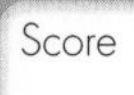

Level ☆

Date / /

Name

Score

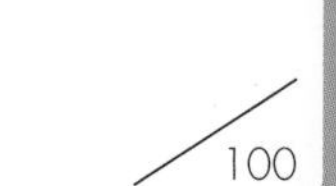

/100

■ The Answer Key is on page 184.

1 Subtract.

3 points per question

(1) $\frac{7}{8}-\frac{1}{2}=$

(2) $\frac{11}{12}-\frac{3}{4}=$

(3) $\frac{13}{14}-\frac{1}{2}=$

(4) $\frac{3}{5}-\frac{2}{7}=$

(5) $\frac{3}{4}-\frac{5}{9}=$

(6) $\frac{3}{4}-\frac{7}{10}=$

(7) $\frac{7}{8}-\frac{1}{6}=$

(8) $\frac{7}{10}-\frac{3}{8}=$

(9) $\frac{11}{20}-\frac{5}{12}=$

(10) $\frac{7}{10}-\frac{9}{14}=$

2 Subtract.

5 points per question

(1) $2\frac{9}{10}-\frac{3}{5}=$

(2) $3\frac{2}{3}-\frac{1}{2}=$

(3) $5\frac{4}{5}-2\frac{3}{4}=$

(4) $4\frac{5}{6}-1\frac{3}{10}=$

(5) $1-\frac{4}{5}=\frac{\square}{5}-\frac{4}{5}=$

(6) $6-\frac{3}{4}=5\frac{\square}{4}-\frac{3}{4}=$

(7) $5-\frac{1}{2}=$

(8) $8\frac{1}{5}-\frac{4}{5}=$

(9) $2\frac{1}{6}-1\frac{8}{9}=$

(10) $\frac{3}{4}-\frac{1}{2}-\frac{1}{5}=$

(11) $\frac{11}{15}-\frac{1}{6}-\frac{2}{5}=$

(12) $5\frac{2}{9}-\frac{5}{12}-\frac{1}{4}=$

3 Answer each word problem. Write the question as an expression first, and then calculate.

10 points for completion

Deborah baked $12\frac{1}{2}$ pounds of cookies. She sold $2\frac{2}{3}$ pounds of cookies in the morning and then sold $5\frac{3}{4}$ pounds in the afternoon. How many pounds of cookies did she have left?

〈Ans.〉

You are doing great!
Keep it up!

5 Multiplication of Fractions Review

Level 

Date / / Name

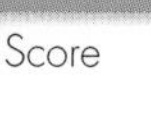

Score /100

■ The Answer Key is on page 184.

1 **Multiply. Reduce as you multiply.**

5 points per question

(1) $\frac{1}{4} \times \frac{3}{5} =$

(2) $\frac{2}{3} \times \frac{4}{7} =$

(3) $\frac{2}{3} \times \frac{7}{8} =$

(4) $\frac{3}{5} \times \frac{4}{21} =$

(5) $\frac{3}{20} \times \frac{5}{6} =$

(6) $\frac{18}{25} \times \frac{10}{21} =$

(7) $1\frac{2}{3} \times \frac{11}{35} =$

(8) $2\frac{1}{4} \times \frac{8}{21} =$

(9) $1\frac{4}{5} \times 10 =$

(10) $2\frac{2}{3} \times 2\frac{1}{10} =$

Convert mixed numbers to improper fractions in order to multiply or divide.

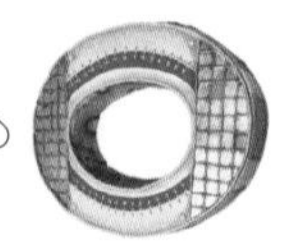

2 Multiply.

5 points per question

(1) $\frac{2}{3} \times \frac{1}{3} \times \frac{4}{5} =$

(2) $\frac{3}{5} \times \frac{1}{9} \times \frac{10}{11} =$

(3) $\frac{8}{15} \times 1\frac{3}{4} \times \frac{10}{21} =$

(4) $3\frac{1}{2} \times \frac{1}{3} \times 1\frac{1}{7} =$

(5) $2\frac{4}{5} \times \frac{3}{8} \times 1\frac{3}{7} =$

(6) $3\frac{3}{4} \times 2\frac{2}{3} \times 2 =$

(7) $5\frac{2}{5} \times \frac{3}{4} \times 1\frac{1}{3} \times \frac{5}{6} =$

(8) $3 \times 1\frac{5}{6} \times \frac{4}{9} \times 2\frac{1}{10} =$

3 Answer each word problem.

5 points per question

(1) Lydia uses $3\frac{1}{5}$ pounds of clay each week in art class. How much clay will she use in $1\frac{3}{4}$ weeks?

〈Ans.〉

(2) A restaurant sells 12 pieces of lasagna each hour during lunch.
How many pieces of lasagna will it sell after $1\frac{5}{6}$ hours?

〈Ans.〉

6 Division of Fractions Review

Level ☆

Date / /

Name

Score /100

■ The Answer Key is on page 185.

1 Divide.

5 points per question

(1) $\frac{1}{2} \div \frac{3}{5} =$

(2) $\frac{2}{3} \div \frac{5}{6} =$

(3) $\frac{4}{15} \div \frac{8}{9} =$

(4) $\frac{8}{21} \div \frac{10}{27} =$

(5) $\frac{20}{27} \div \frac{10}{33} =$

(6) $\frac{2}{3} \div 8 =$

(7) $6 \div \frac{3}{7} =$

(8) $\frac{12}{25} \div 2\frac{4}{5} =$

(9) $1\frac{1}{3} \div \frac{16}{21} =$

(10) $4\frac{2}{7} \div 2\frac{4}{7} =$

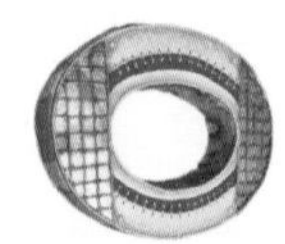

Reduce as you calculate.

2 Divide.

5 points per question

(1) $\frac{1}{3} \div \frac{5}{7} \div \frac{9}{10} =$

(2) $\frac{2}{5} \div \frac{3}{7} \div \frac{21}{40} =$

(3) $\frac{5}{18} \div \frac{20}{27} \div \frac{1}{4} =$

(4) $\frac{5}{8} \div 3 \div \frac{7}{12} =$

(5) $1\frac{1}{2} \div \frac{15}{16} \div 2\frac{2}{3} =$

(6) $3\frac{3}{5} \div 3\frac{1}{3} \div 1\frac{1}{3} =$

(7) $2\frac{3}{4} \div 3\frac{1}{3} \div 4\frac{2}{5} \div 1\frac{1}{8} =$

(8) $\frac{4}{7} \div 1\frac{5}{6} \div 3 \div 1\frac{1}{7} =$

3 Answer each word problem. Write the question as an expression first, and then calculate.

5 points per question

(1) An athlete runs $3\frac{1}{2}$ kilometers every $10\frac{1}{4}$ minutes. How far does the athlete run each minute?

〈Ans.〉

(2) A box of paper weighs $6\frac{2}{3}$ ounces. Each box contains $1\frac{1}{3}$ dozen sheets of paper. How much does each sheet of paper weigh?

〈Ans.〉

7 Decimals and Fractions

Level 

Date / /

Name

Score /100

■ The Answer Key is on page 185.

1 **Rewrite each decimal as a fraction. Then calculate.**

5 points per question

(1) $0.5+\frac{1}{7}=$

(6) $\frac{3}{5}-0.25=$

(2) $1.2+\frac{2}{3}=$

(7) $6.8-2\frac{1}{3}=$

(3) $4\frac{3}{4}+3.25=$

(8) $4.125-1\frac{3}{4}=$

(4) $5\frac{1}{6}+2.125=$

(9) $7\frac{1}{2}-3.75=$

(5) $1.05+3\frac{3}{8}+2.5=$

(10) $\frac{9}{10}-0.5-0.375=$

2 Multiply or divide.

5 points per question

(1) $\frac{1}{3} \times 0.6 =$

(2) $1.25 \times \frac{8}{15} =$

(3) $3.125 \times 3\frac{1}{5} =$

(4) $3\frac{1}{3} \times 2.2 =$

(5) $2.8 \times 6.25 \times \frac{3}{7}$

$=$

(6) $1.5 \div \frac{9}{10} =$

(7) $\frac{5}{8} \div 1.125 =$

(8) $2\frac{1}{5} \div 1.7 =$

(9) $3.375 \div 2\frac{1}{4} =$

(10) $2\frac{2}{3} \div 1.25 \div 0.8$

$=$

Impressive!

8 Exponent Review

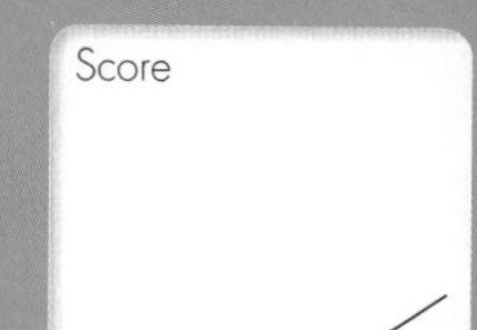

Date / / Name

■ The Answer Key is on page 185.

1 Calculate.

5 points per question

(1) $2^3 =$

(2) $3^2 =$

(3) $2^1 =$

(4) $3^3 =$

(5) $2^5 =$

(6) $2^0 =$

(7) $3^0 =$

(8) $6^3 =$

(9) $7^3 =$

(10) $2^7 =$

Any number raised to the power of 0 equals 1.

2 Calculate.

5 points per question

(1) $\left(\frac{1}{2}\right)^3 =$

(2) $\left(\frac{1}{4}\right)^2 =$

(3) $\left(\frac{2}{3}\right)^4 =$

(4) $\left(2\frac{2}{3}\right)^2 =$

(5) $\left(1\frac{1}{3}\right)^3 =$

(6) $\frac{2}{3^3} =$

(7) $\frac{3^4}{5} =$

(8) $\frac{2^3}{3^4} =$

(9) $\frac{3^3}{7^2} =$

(10) $\frac{2^6}{9^3} =$

9 Exponent Review

Level 

Date / /

Name

Score /100

■ The Answer Key is on page 185.

1 **Calculate.**

4 points per question

Example $2^4 \times 2^2 = 2 \times 2 \times 2 \times 2 \times 2 \times 2 = 64$

(1) $3^2 \times 3^3 =$

(2) $4^1 \times 4^2 =$

(3) $2^0 \times 2^4 =$

(4) $2^3 \times 2^5 =$

(5) $5^3 \times 5^1 =$

(6) $\left(\frac{1}{2}\right)^2 \times \left(\frac{1}{2}\right)^3 =$

(7) $\left(\frac{1}{4}\right)^1 \times \left(\frac{1}{4}\right)^2 =$

(8) $\left(\frac{2}{3}\right)^3 \times \left(\frac{2}{3}\right)^1 =$

(9) $\left(\frac{3}{4}\right)^2 \times \left(\frac{3}{4}\right)^2 =$

(10) $\left(\frac{2}{5}\right)^2 \times \left(\frac{2}{5}\right)^3 =$

2 Calculate.

5 points per question

(1) $2^4 \times \left(\frac{1}{2}\right)^2 = \overset{1}{\cancel{2}} \times \overset{\boxed{1}}{\cancel{2}} \times 2 \times 2 \times \frac{1}{\underset{1}{\cancel{2}}} \times \frac{1}{\underset{\boxed{1}}{\cancel{2}}}$

$=$

(2) $5^3 \times \left(\frac{1}{5}\right)^1 =$

(3) $\left(\frac{1}{4}\right)^2 \times 4^5 =$

(4) $\left(\frac{1}{3}\right)^4 \times 3^2 =$

(5) $5^3 \times \left(\frac{2}{5}\right)^1 =$

(6) $6^2 \times \left(\frac{5}{6}\right)^3 =$

(7) $\left(\frac{2}{9}\right)^1 \times \left(\frac{1}{4}\right)^2 =$

(8) $\left(\frac{3}{5}\right)^3 \times \left(\frac{2}{9}\right)^2 =$

(9) $\left(\frac{3}{4}\right)^3 \times \left(\frac{2}{9}\right)^2 =$

(10) $\left(3\frac{1}{2}\right)^1 \times \left(1\frac{1}{3}\right)^2 =$

3 Answer each word problem. Write the question as an expression first, and then calculate.

10 points for completion

The volume of a room can be calculated by multiplying the length, width, and height. If the length and width are both $2\frac{1}{2}$ meters and the height is 6 meters, what is the volume of the room?

⟨Ans.⟩ ______ m^3

You really understand exponents!

Exponents

Multiplication

Date  Name

Level 

Score  100

■ The Answer Key is on page 185.

1 **Multiply. Write the intermediate steps taken to calculate the answer.** 4 points per question

Don't forget!

When multiplying exponents with the same base, add the exponents.

Example $2^4 \times 2^1 = 2^{4+1} = 2^5 = 32$

(1) $3^3 \times 3^2 = 3^5 =$

(2) $4^1 \times 4^2 = 4^3$

(3) $3^0 \times 3^4 =$

(4) $2^3 \times 2^4 =$

(5) $5^2 \times 5^2 =$

(6) $\left(\frac{1}{2}\right)^3 \times \left(\frac{1}{2}\right)^2 =$

(7) $\left(\frac{1}{4}\right)^2 \times \left(\frac{1}{4}\right)^2 =$

(8) $\left(\frac{2}{3}\right)^2 \times \left(\frac{2}{3}\right)^1 =$

(9) $\left(1\frac{3}{4}\right)^1 \times \left(1\frac{3}{4}\right)^1 =$

(10) $\left(2\frac{1}{5}\right)^0 \times \left(2\frac{1}{5}\right)^2 =$

2 Multiply.

5 points per question

(1) $\frac{2^3}{3} \times \frac{2^1}{7} = \frac{2^{\square}}{21} =$

(2) $\frac{4^1}{7} \times \frac{4^2}{11} =$

(3) $\frac{2}{3^2} \times \frac{4}{3^3} =$

(4) $\frac{5}{2^3} \times \frac{3}{2^2} =$

(5) $\frac{3}{2^0} \times \frac{4}{2^1} =$

(6) $\frac{2^5}{3^2} \times \frac{2^1}{3^3} =$

(7) $\frac{4^1}{5^2} \times \frac{4^2}{5^1} =$

(8) $\frac{3^3}{6^0} \times \frac{3^2}{6^3} =$

(9) $\frac{2^4}{7^1} \times \frac{2^2}{7^1} =$

(10) $\frac{3^1}{4^2} \times \frac{3^3}{4^2} =$

(11) $\frac{3^2}{2^2} \times \frac{1}{2^3} \times \frac{3^1}{5} =$

(12) $\frac{2}{5^0} \times \frac{4^1}{5^2} \times \frac{4^1}{5^1} =$

Exponents
Division

Date / /

Name

Level

Score /100

■ The Answer Key is on page 185.

Divide.

5 points per question

Don't forget!

When dividing exponents with the same base, subtract the exponents.

Example $2^4 \div 2^1 = 2^{4-1} = 2^3 = 8$

(1) $4^3 \div 4^1 = 4^2 =$

(2) $3^5 \div 3^3 =$

(3) $5^4 \div 5^1 =$

(4) $2^6 \div 2^2 =$

(5) $7^8 \div 7^6 =$

(6) $\left(\frac{1}{3}\right)^5 \div \left(\frac{1}{3}\right)^1 =$

(7) $\left(\frac{1}{2}\right)^9 \div \left(\frac{1}{2}\right)^5 =$

(8) $\left(\frac{3}{4}\right)^4 \div \left(\frac{3}{4}\right)^3 =$

(9) $\left(\frac{2}{5}\right)^3 \div \left(\frac{2}{5}\right)^0 =$

(10) $\left(2\frac{3}{4}\right)^5 \div \left(2\frac{3}{4}\right)^3 =$

2 Calculate.

5 points per question

(1) $2^3 \times 2^6 \div 2^4 = 2^{3+6-4} =$

(2) $3^1 \times 3^5 \div 3^2 =$

(3) $2^7 \div 2^4 \times 2^1 =$

(4) $\left(\frac{1}{2}\right)^7 \div \left(\frac{1}{2}\right)^2 \div \left(\frac{1}{2}\right)^2 =$

(5) $\left(\frac{2}{3}\right)^8 \div \left(\frac{2}{3}\right)^5 \div \left(\frac{2}{3}\right)^2 =$

(6) $\left(\frac{2}{5}\right)^9 \div \left(\frac{2}{5}\right)^7 \times \left(\frac{2}{5}\right)^1 =$

(7) $\left(3\frac{2}{3}\right)^9 \times \left(3\frac{2}{3}\right)^1 \div \left(3\frac{2}{3}\right)^8 =$

(8) $\left(4\frac{1}{2}\right)^6 \div \left(4\frac{1}{2}\right)^5 \times \left(4\frac{1}{2}\right)^0 =$

(9) $\left(3\frac{3}{5}\right)^1 \times \left(3\frac{3}{5}\right)^2 \div \left(3\frac{3}{5}\right)^3 =$

(10) $\left(2\frac{1}{5}\right)^5 \div \left(2\frac{1}{5}\right)^1 \div \left(2\frac{1}{5}\right)^2 =$

Exponents
Division

Date / /

Name

Level 

Score /100

■ The Answer Key is on page 186.

1 **Divide.**

5 points per question

Don't forget!

When dividing numbers with the same exponents, simplify whenever possible.

Example $6^4 \div 2^4 = \left(\frac{\overset{3}{\not{6}}}{\underset{1}{\not{2}}}\right)^4 = 3^4 = 81$

(1) $12^2 \div 4^2 = \left(\frac{12}{4}\right)^2 =$

(2) $6^3 \div 3^3 = \left(\frac{6}{3}\right)^3 =$

(3) $15^5 \div 5^5 = \left(\frac{15}{5}\right)^5 =$

(4) $100^4 \div 10^4 =$

(5) $56^2 \div 7^2 =$

(6) $2^3 \div 8^3 =$

(7) $6^4 \div 18^4 =$

(8) $10^3 \div 25^3 =$

(9) $18^3 \div 12^3 =$

(10) $30^3 \div 24^3 =$

In division, you can simplify only when the exponents are the same.

2 **Divide.**

5 points per question

(1) $\left(\frac{1}{3}\right)^3 \div (2)^3 = \left(\frac{1}{3} \div 2\right)^3$

$= \left(\frac{1}{3} \times \frac{1}{2}\right)^3 =$

(2) $\left(\frac{2}{5}\right)^2 \div \left(\frac{4}{7}\right)^2 =$

(3) $\left(\frac{3}{4}\right)^2 \div 4^2 =$

(4) $3^2 \div \left(\frac{6}{7}\right)^2 =$

(5) $\left(\frac{1}{2}\right)^3 \div \left(\frac{3}{8}\right)^3 =$

(6) $\left(\frac{2}{3}\right)^3 \div \left(\frac{8}{9}\right)^3 =$

(7) $\left(\frac{4}{9}\right)^2 \div \left(\frac{8}{15}\right)^2 =$

(8) $\left(2\frac{2}{3}\right)^2 \div \left(1\frac{1}{9}\right)^2 =$

(9) $\left(4\frac{1}{2}\right)^2 \div \left(1\frac{1}{5}\right)^2 =$

(10) $\left(\frac{2}{3}\right)^3 \div \left(1\frac{1}{9}\right)^3 \div 6^3 =$

You are getting exponentially smarter!

13 Exponents Review

Level 

Date / / Name

Score /100

■ The Answer Key is on page 186.

1 Calculate.

4 points per question

(1) $3^6 \times \frac{1}{3^4} =$

(6) $2^4 \times 2^3 =$

(2) $\left(\frac{3}{8}\right)^3 \times 4^2 =$

(7) $\left(\frac{1}{3}\right)^2 \times \left(\frac{1}{3}\right)^3 =$

(3) $\left(\frac{2}{3}\right)^3 \times \left(\frac{3}{4}\right)^2 =$

(8) $2^0 \times 2^1 \times 2^2 \times 2^3 =$

(4) $\left(\frac{4}{9}\right)^2 \times \left(\frac{15}{16}\right)^2 =$

(9) $\frac{2^3}{5^2} \times \frac{2^4}{5^1} =$

(5) $\left(2\frac{2}{5}\right)^2 \times \left(2\frac{1}{2}\right)^4 =$

(10) $\frac{2^1}{5} \times \frac{2^3}{3} \times \frac{2^2}{3^3} =$

Reduce as you calculate.

2 Calculate.

5 points per question

(1) $3^8 \div 3^4 \div 3^2 =$

(2) $\left(\frac{3}{5}\right)^9 \div \left(\frac{3}{5}\right)^2 \div \left(\frac{3}{5}\right)^4 =$

(3) $\left(2\frac{1}{2}\right)^{10} \div \left(2\frac{1}{2}\right)^5 \div \left(2\frac{1}{2}\right)^2$

$=$

(4) $3^4 \times 3^6 \div 3^7 =$

(5) $\left(2\frac{1}{3}\right)^6 \div \left(2\frac{1}{3}\right)^5 \times \left(2\frac{1}{3}\right)^2 \div \left(2\frac{1}{3}\right)^1$

$=$

(6) $15^4 \div 5^4 =$

(7) $8^3 \div 20^3 =$

(8) $35^3 \div 21^3 =$

(9) $\left(\frac{2}{3}\right)^3 \div 6^3 =$

(10) $\left(\frac{3}{5}\right)^2 \div \left(\frac{9}{20}\right)^2 =$

(11) $\left(3\frac{1}{3}\right)^2 \div \left(4\frac{2}{3}\right)^2 =$

(12) $\left(2\frac{2}{5}\right)^2 \times \left(3\frac{1}{3}\right)^2 \div \left(\frac{2}{3}\right)^2$

$=$

Order of Operations Review

Level ★

Date / /

Name

Score /100

■ The Answer Key is on page 186.

1 Calculate. Write the intermediate steps.

3 points per question

Don't forget!

According to the order of operations, calculate from left to right.

Examples $7-4+5=3+5=8$ $12\div2\times3=6\times3=18$

(1) $10+5-3=$

(2) $7-3+5=$

(3) $10-5+8=$

(4) $\frac{1}{2}+3-2=$

(5) $1\frac{1}{4}+\frac{1}{2}-1\frac{1}{3}=$

(6) $16\div4\times3=$

(7) $10\times3\div5=$

(8) $9\times3\div6=$

(9) $\frac{1}{2}\div\frac{1}{3}\times\frac{1}{9}=$

(10) $2\frac{1}{2}\div\frac{3}{4}\times\frac{1}{6}=$

2 Calculate. Write the intermediate steps.

5 points per question

Don't forget!

According to the order of operations,

- **perform multiplication and division before addition and subtraction**
- then calculate from left to right

Examples $10 \div 2 + 3 \times 2 = 5 + 6 = 11$ $5 + 3 \times 2 - 4 = 5 + 6 - 4 = 7$

(1) $3 + 2 \times 4 = 3 + \square =$

(2) $8 \div 4 + 5 = \square + 5 =$

(3) $9 \times 2 - 10 =$

(4) $\frac{1}{2} + 5 \div 2 =$

(5) $8 - 3 \times \frac{1}{6} =$

(6) $3 + 9 + 1 \times 3 =$

(7) $12 \div 2 - 1 - 5 =$

(8) $7\frac{1}{2} - 6 \div 2 + \frac{1}{4} =$

(9) $\frac{1}{2} \times \frac{4}{5} + 1 - \frac{2}{3} =$

(10) $2 \times 3 + 1 \times 5 =$

(11) $3 \times \frac{1}{2} + 6 \div \frac{1}{2} =$

(12) $\frac{5}{6} \div \frac{1}{3} - \frac{2}{5} \div \frac{2}{3} =$

3 Answer each word problem. Write the question as an expression first, and then calculate.

10 points for completion

Samuel is counting the number of animals in a field. In the first 4 hours, he counts 5 elephants each hour. In the next 2 hours, he counts 12 zebras each hour. How many animals did Samuel count altogether?

〈**Ans.**〉

Order of Operations

Level 

Date / /

Name

Score /100

■ The Answer Key is on page 186.

1 Calculate. Write the intermediate steps.

3 points per question

Don't forget!

According to the order of operations,

- **calculate the numbers in parentheses and brackets first**
- then perform multiplication and division before addition and subtraction
- then calculate from left to right

(1) $9-3+5=$

(2) $9-(3+5)=9-\square=$

(3) $6\times2\div4=$

(4) $(6\times2)\div4=$

(5) $6\times(2\div4)=$

(6) $7-2\times2-1=$

(7) $(7-2)\times(2-1)=$

(8) $(7-2)\times2-1=$

(9) $7-2\times(2-1)=$

(10) $7-(2\times2-1)=$

2 Calculate. Write the intermediate steps.

5 points per question

(1) $2-\frac{1}{2}+3\times\frac{1}{3}=$

(2) $\left(2-\frac{1}{2}+3\right)\times\frac{1}{3}=$

(3) $2-\left(\frac{1}{2}+3\right)\times\frac{1}{3}=$

(4) $20-[4\div(2+6)]\div 2$

$=$

(5) $[(20-4)\div(2+6)]\div 10$

$=$

(6) $1\frac{3}{4}-\frac{1}{2}\div 4\times 2=$

(7) $1\frac{3}{4}-\frac{1}{2}\div(4\times 2)=$

(8) $\left(1\frac{3}{4}-\frac{1}{2}\div 4\right)\times 2=$

(9) $\left[\left(1\frac{3}{4}-\frac{1}{2}\right)\div 4\right]\times 2=$

(10) $\left(1\frac{3}{4}-\frac{1}{2}\right)\div(4\times 2)=$

(11) $8.2-\frac{1}{5}+3\div 6\times 2$

$=$

(12) $\left(8.2-\frac{1}{5}\right)+[3\div(6\times 2)]$

$=$

3 Answer each word problem. Write the question as an expression first, and then calculate.

10 points for completion

Robert bought 5 books each day for 2 days. He then bought 2 books each day for 4 days. If each book is worth $15.00, how much money did he spend?

〈Ans.〉 ____________

16 Order of Operations

Date / /

Name

Score /100

■ The Answer Key is on page 187.

1 **Calculate. Write the intermediate steps.**

6 points per question

Don't forget!

According to the order of operations,

- **calculate exponents and numbers in parentheses and brackets first**
- perform multiplication and division before addition and subtraction
- calculate from left to right

(1) $5+2^3\times4=5+\square\times4=$

(2) $(5+2^3)\times4=(5+\square)\times4$

$=\square\times4=$

(3) $6^2-2^5\times6^0+\left(\frac{1}{2}\right)^3$

$=$

(4) $\left(\frac{1}{2}\right)^2\times3+5\div2=$

(5) $4-\left(\frac{2}{3}\right)^3\times12=$

(6) $\frac{2^3}{3^2}\times(12-9)\div\frac{10}{27}=$

(7) $5+(1+3)^2\div20=$

(8) $(4\div6)^2\times18-4=$

(9) $\frac{(8-3\times2)^2}{1+2\times3}=$

(10) $\left(9\div\frac{1}{2}-8\times2\right)+(3^2-2^3)^2\times\left(\frac{1}{3}\right)^2$

$=$

2 Calculate. Write the intermediate steps.

5 points per question

(1) $19-2^{3+1}=19-2^{\square}=$

(2) $3^{2\times2}\div2^{3+1}=$

(3) $2^{6\div3}+\left(\frac{1}{2}\right)^{5-2}=$

(4) $\left[\left(2.75-\frac{7}{8}\right)\div\frac{5}{16}\right]^{7-4}$

$=$

(5) $3.4+6\times\left[2-\left(\frac{1}{2}\right)^{1+2}\right]$

$=$

(6) $12\div(15-11)^{6\div3}+\left(3.5-\frac{3}{4}\right)$

$=$

(7) $\left(\frac{1}{2}+\frac{5}{6}\right)^{6\div3}-1\div\left(0.8+\frac{1}{10}\right)$

$=$

(8) $\frac{(9-4)^{8-6}}{3^2\times2^{6-3}}=$

17 Order of Operations Review

Level 

Date / /

Name

Score /100

■ The Answer Key is on page 187.

1 Calculate.

5 points per question

(1) $5 \div (3-2) \times 6 =$

(2) $5 \div (3-2) \times (6+2) =$

(3) $5 \div [(3-2) \times (6+2)] =$

(4) $\frac{1}{4} + \frac{1}{2} \div \frac{1}{3} \times \frac{1}{5} =$

(5) $\frac{1}{4} + \frac{1}{2} \div \left(\frac{1}{3} \times \frac{1}{5}\right) =$

(6) $\left(\frac{1}{4} + \frac{1}{2}\right) \div \frac{1}{3} \times \frac{1}{5} =$

(7) $\left(\frac{1}{4} + \frac{1}{2} \div \frac{1}{3}\right) \times \frac{1}{5} =$

(8) $1 + 2 \times 3 - 4 \div 5 =$

(9) $(1+2) \times 3 - 4 \div 5 =$

(10) $[(1+2) \times 3 - 4] \div 5 =$

If you find this page difficult, please review pages 28 to 31.

2 Calculate.

5 points per question

(1) $0.75+\frac{3}{4}\times\frac{1}{2}-\left(\frac{1}{3}\right)^2$

$=$

(2) $0.75+\frac{3}{4}\times\left(\frac{1}{2}-\frac{1}{3}\right)^2$

$=$

(3) $\left(0.75+\frac{3}{4}\right)^2\times\left(\frac{1}{2}-\frac{1}{3}\right)$

$=$

(4) $0.75+\left(\frac{3}{4}\right)^2\times\frac{1}{2}-\frac{1}{3}$

$=$

(5) $\left(0.75+\frac{3}{4}\right)^2\times\frac{1}{2}-\frac{1}{3}$

$=$

(6) $3+1\frac{1}{2}\times1\frac{1}{3}+2^2\times3$

$=$

(7) $3+1\frac{1}{2}\times\left(1\frac{1}{3}+2\right)^2\times3$

$=$

(8) $\left[3+\left(1\frac{1}{2}\times1\frac{1}{3}+2\right)^2\right]\times3$

$=$

(9) $\left[3+\left(1\frac{1}{2}\times1\frac{1}{3}+2\right)\right]^2\times3$

$=$

(10) $\left[\left(3+1\frac{1}{2}\right)\times\left(1\frac{1}{3}+2\right)\right]^2\times3$

$=$

If you find this page difficult, please review pages 30 to 33.

You figured it out!

Negative Numbers

Date / /

Name

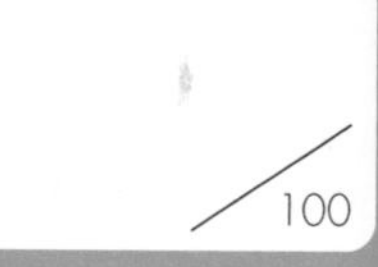

■ The Answer Key is on page 187.

1 Calculate.

2 points per question

Examples

3−0=3

3−1=2

3−2=1

3−3=0

3−4=−1

3−5=−2

Negative numbers are real numbers that are less than 0. −1 is read as "negative one." −2 is read as "negative two."

(1) 4−2=

(2) 4−3=

(3) 4−4=

(4) 4−5=−1

(5) 4−6=−2

(6) 6−4=

(7) 6−6=

(8) 6−7=

(9) 6−9=

(10) 6−10=

Both positive and negative numbers may be graphed on a number line.

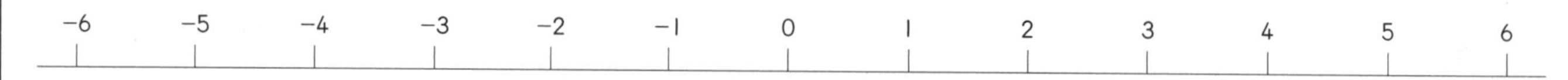

2 Calculate.

5 points per question

(1) $8-7=$

(2) $8-8=$

(3) $8-9=$

(4) $8-10=$

(5) $10-8=$

(6) $10-9=$

(7) $10-10=$

(8) $10-11=$

(9) $15-12=$

(10) $15-14=$

(11) $15-15=$

(12) $15-16=$

(13) $15-17=$

(14) $15-20=$

(15) $15-25=$

(16) $15-40=$

Don't forget!

Numbers greater than 0, such as 1, 2, 10, 3.54, and $\frac{1}{2}$ are called **positive numbers**. Numbers less than 0 are called **negative numbers** and have a **negative sign** (–) before them, such as -1, -2, -3.5, and $-4\frac{1}{2}$. The number 0 is a special number and is neither positive nor negative.

19 Addition with Negative Numbers

Level 

Date / /　Name

Score /100

■ The Answer Key is on page 187.

1 Add.

2 points per question

Examples $-2+3=1$

$-3+3=0$

$-4+3=-1$

(1) $1+2=$

(2) $0+2=$

(3) $-1+2=$

(4) $-2+2=$

(5) $-3+2=-1$

(6) $6+4=$

(7) $0+4=$

(8) $-3+4=$

(9) $-4+4=$

(10) $-9+4=$

2 Add.

4 points per question

(1) $3+6=$

(2) $-2+6=$

(3) $0+6=$

(4) $-7+6=$

(5) $-5+6=$

(6) $1+6=$

(7) $-1+3=$

(8) $-7+3=$

(9) $-3+3=$

(10) $-2+3=$

(11) $-6+8=$

(12) $3+0=$

(13) $-5+1=$

(14) $-6+5=$

(15) $0+4=$

(16) $-9+4=$

(17) $-7+7=$

(18) $-8+20=$

(19) $-20+100=$

(20) $-100+300=$

You've got it!

20 Subtraction with Negative Numbers

Level ☆☆

Date / /

Name

Score /100

■ The Answer Key is on page 187.

1 Subtract.

2 points per question

Examples $2-1=1$

$1-1=0$

$0-1=-1$

$-1-1=-2$

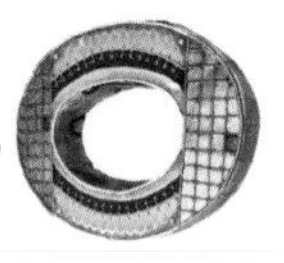

If you find this page difficult, refer to the number line on page 36.

(1) $3-2=$

(2) $2-2=$

(3) $1-2=$

(4) $0-2=$

(5) $-1-2=$

(6) $3-4=$

(7) $-3-4=$

(8) $6-4=$

(9) $-6-4=$

(10) $1-4=$

(11) $-1-4=$

(12) $2-4=$

(13) $-2-4=$

(14) $-4-9=$

(15) $-8-5=$

(16) $-10-4=$

(17) $-7-2=$

(18) $-16-9=$

(19) $-10-20=$

(20) $-12-13=$

2 Subtract.

4 points per question

(1) $\frac{2}{4}-\frac{1}{4}=$

(2) $\frac{1}{4}-\frac{1}{4}=$

(3) $0-\frac{1}{4}=-\frac{\square}{4}$

(4) $-\frac{1}{4}-\frac{1}{4}=-\frac{\square}{4}=$

(5) $-\frac{2}{4}-\frac{1}{4}=$

(6) $\frac{2}{5}-\frac{3}{5}=$

(7) $-\frac{3}{5}-\frac{3}{5}=$

(8) $-2\frac{1}{5}-\frac{3}{5}=-2\frac{4}{5}$

(9) $\frac{1}{4}-\frac{1}{2}=\frac{1}{4}-\frac{\square}{4}=$

(10) $-\frac{1}{4}-\frac{1}{6}=$

(11) $1-4\frac{2}{3}=-3\frac{2}{3}$

(12) $\frac{3}{10}-\frac{7}{15}=$

(13) $1\frac{1}{4}-\frac{1}{2}=$

(14) $\frac{1}{2}-1\frac{1}{4}=\frac{\square}{4}-1\frac{1}{4}=\frac{2}{4}-\frac{\square}{4}=$

(15) $1\frac{1}{6}-5\frac{2}{3}=$

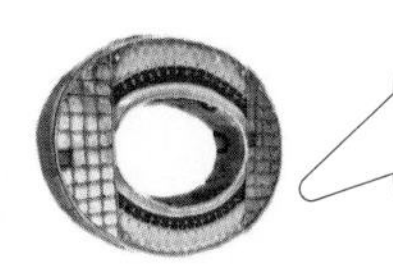

Don't forget to reduce!

Super job!

21 Subtraction with Negative Numbers

Date / /

Name

Score /100

■ The Answer Key is on page 188.

Write + or − in the boxes. Then calculate.

2 points per question

Addition examples

$5+(-3)=5-3=2$

$-1+(-7)=-1-7=-8$

Subtraction examples

$2-6=-4$

$3-(-4)=3+4=7$

$-5-(-1)=-5+1=-4$

(1) $4+(-1)=4\square 1=$

(2) $-2+(-4)=-2\square 4=$

(3) $0+(-5)=$

(4) $\frac{1}{2}+\left(-\frac{1}{3}\right)=$

(5) $-3\frac{1}{4}+\left(-\frac{2}{3}\right)=$

(6) $2-(-6)=2\square 6=$

(7) $-3-(-2)=-3\square 2=$

(8) $0-(-4)=$

(9) $\frac{2}{3}-\left(-\frac{1}{3}\right)=$

(10) $\frac{2}{5}-\left(-1\frac{1}{5}\right)=$

2 **Calculate.**

4 points per question

(1) $5+2=$

(2) $5-2=$

(3) $-5+2=$

(4) $-5-2=$

(5) $5-(-2)=$

(6) $5+(-2)=$

(7) $-5-(-2)=$

(8) $-5+(-2)=$

(9) $3-(-6)=$

(10) $-3+(-6)=$

(11) $\frac{2}{3}-\frac{1}{3}=$

(12) $\frac{2}{3}-\left(-\frac{1}{3}\right)=$

(13) $-\frac{2}{3}-\left(-\frac{1}{3}\right)=$

(14) $-\frac{2}{3}+\left(-\frac{1}{3}\right)=$

(15) $\frac{1}{3}-\frac{2}{3}=$

(16) $\frac{1}{3}-\left(-\frac{2}{3}\right)=$

(17) $-\frac{1}{3}+\left(-\frac{2}{3}\right)=$

(18) $-\frac{1}{3}-\left(-\frac{2}{3}\right)=$

(19) $\frac{1}{2}-\left(-1\frac{1}{3}\right)=$

(20) $\frac{1}{2}+\left(-1\frac{1}{3}\right)=$

Your work is really adding up!

22 Multiplication with Negative Numbers

Level ★★

Date / /

Name

Score /100

■ The Answer Key is on page 188.

1 **State the number of negative signs. Circle whether the answer is positive or negative.**

3 points per question

Don't forget!

When multiplying negative numbers, first count the number of negative signs.

- If there is an even number of negative signs, the answer is positive.
- If there is an odd number of negative signs, the answer is negative.

Example $2 \times (-1) \times 3 \times (-5)$ 2 negative signs → positive answer

(1) $3 \times (-1) \times (-2)$

☐ negative signs

positive / negative answer

(2) $1 \times (-4) \times (-3)$

☐ negative signs

positive / negative answer

(3) $(-2) \times 6 \times (-1) \times (-3)$

☐ negative signs

positive / negative answer

(4) $(-5) \times (-4) \times (-3) \times 2 \times (-1)$

☐ negative signs

positive / negative answer

2 **Determine the sign of the answer. Then multiply.**

4 points per question

Don't forget!

After determining if the answer is positive or negative, multiply the numbers to find the answer.

Example $3 \times (-2) \times (-1) \times (-2) = -(3 \times 2 \times 1 \times 2) = -12$

(1) $(-2) \times (-1) \times (-3) =$

(2) $5 \times (-3) \times (-1) =$

(3) $2 \times 3 \times (-5) =$

(4) $(-1) \times 2 \times (-3) \times 4 =$

(5) $3 \times 0 \times (-2) \times (-3) =$

(6) $(-1) \times 1 \times (-1) \times 1 \times (-1) =$

3 Determine the sign of the answer. Then multiply.

4 points per question

(1) $(-1)\times(-1)\times(-1)=$

(2) $(-1)\times1\times1\times(-1)\times1=$

(3) $(-1)\times1\times(-1)\times1\times(-1)\times1=$

(4) $3\times(-4)\times1=$

(5) $(-2)\times(-3)\times6=$

(6) $4\times3\times\left(-\frac{1}{2}\right)=$

(7) $\left(-\frac{1}{3}\right)\times(-8)\times\left(-\frac{9}{10}\right)=$

(8) $1\frac{1}{3}\times(-4)\times\left(-\frac{9}{10}\right)=$

(9) $\left(-3\frac{1}{4}\right)\times\frac{2}{13}\times\left(-2\frac{1}{6}\right)=$

(10) $\left(-2\frac{1}{2}\right)\times\left(-1\frac{1}{5}\right)\times1\frac{3}{4}$

$=$

(11) $(-3)(-2)(-5)=$

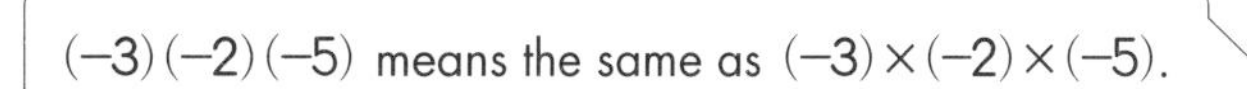
(−3)(−2)(−5) means the same as (−3)×(−2)×(−5).

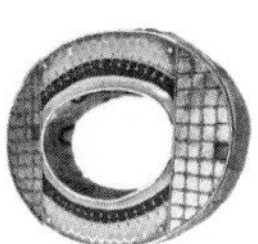

(12) $(-4)(-1)\times2\times(-3)=$

(13) $2\times(-4)(-1)\times2=$

(14) $\left(-1\frac{1}{2}\right)\left(-3\frac{1}{3}\right)\times\frac{1}{4}\times\left(-1\frac{1}{5}\right)$

$=$

(15) $\frac{3}{5}\times\left(-1\frac{1}{6}\right)\left(-\frac{2}{5}\right)\times1\frac{1}{9}$

$=$

(16) $1.2\times\frac{3}{4}\times(-0.6)\times1\frac{1}{9}$

$=$

Multiply your efforts!

23 Division with Negative Numbers

Level ★★

Date / /

Name

Score /100

■ The Answer Key is on page 188.

1 **Determine the sign of the answer. Then divide.** 2 points per question

Don't forget!

Before dividing negative numbers, count the number of negative signs to determine if the answer is positive or negative. Then calculate the answer.

Examples $6 \div (-3) = -2$ $\quad -24 \div 4 \div (-2) = 24 \div 4 \div 2 = 3$

(1) $(-10) \div 2 =$

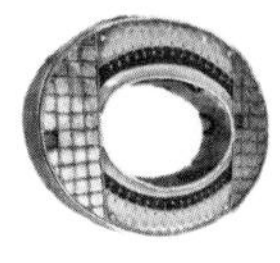

(−10) is the same as −10.

(2) $-8 \div (-4) =$

(3) $0 \div (-3) =$

(4) $16 \div (-8) =$

(5) $(-20) \div (-5) =$

(6) $24 \div (-3) \div 4 =$

(7) $30 \div 5 \div (-6) =$

(8) $-60 \div 2 \div (-10) =$

(9) $80 \div (-4) \div (-5) \div (-2) =$

(10) $120 \div (-3) \div 2 \div 4 =$

2 Determine the sign of the answer. Then divide.

4 points per question

(1) $8 \div \left(-\frac{1}{2}\right) =$

(2) $(-9) \div \frac{3}{4} =$

(3) $-\frac{1}{6} \div \left(-\frac{1}{2}\right) =$

(4) $\frac{2}{3} \div \left(-\frac{4}{15}\right) =$

(5) $-\frac{4}{9} \div \frac{6}{5} =$

(6) $\left(-1\frac{2}{3}\right) \div \left(-\frac{1}{6}\right) =$

(7) $\frac{5}{8} \div \left(-2\frac{1}{2}\right) =$

(8) $-3\frac{1}{3} \div \frac{2}{9} =$

(9) $-4\frac{1}{2} \div \left(-2\frac{2}{5}\right) =$

(10) $\left(-2\frac{1}{4}\right) \div 4\frac{1}{2} =$

(11) $0.5 \div \left(-\frac{3}{4}\right) =$

(12) $-2\frac{1}{5} \div (-0.8) =$

(13) $12 \div \frac{1}{2} \div (-3) =$

(14) $(-6) \div \frac{1}{4} \div (-8) =$

(15) $-\frac{1}{3} \div \left(-\frac{3}{4}\right) \div \left(-\frac{5}{6}\right) =$

(16) $9 \div \left(-1\frac{4}{5}\right) \div \frac{5}{6} =$

(17) $-\frac{1}{6} \div \left(-1\frac{2}{3}\right) \div 4 =$

(18) $3\frac{3}{4} \div (-5) \div \left(-1\frac{1}{5}\right) =$

(19) $-2.4 \div \left(-1\frac{1}{2}\right) \div (-0.9) =$

(20) $\left(-1\frac{1}{2}\right) \div 0.6 \div (-0.25) =$

24

Multiplication & Division with Negative Numbers

Level 

Date / /

Name

Score /100

■ The Answer Key is on page 188.

1 **Determine the sign of the answer. Then calculate.**

3 points per question

(1) $-30 \div (-5) \div (-6) =$

(2) $-30 \div [(-5) \div (-6)] =$

(3) $10 \times 0 \div (-3) =$

(4) $-3 \div (-4) \times \left(-\frac{4}{5}\right) =$

(5) $-3 \div \left[(-4) \times \left(\frac{4}{5}\right)\right] =$

(6) $\frac{1}{5} \times (-3) \div \frac{6}{11} =$

(7) $-\frac{3}{5} \div \left(-\frac{9}{10}\right) \div (-6) =$

(8) $-\frac{3}{5} \div \left[\left(-\frac{9}{10}\right) \div (-6)\right] =$

(9) $-1\frac{2}{3} \div \frac{5}{6} \times \left(-2\frac{1}{2}\right) =$

(10) $2\frac{1}{4} \times \left(-3\frac{1}{3}\right) \div 1\frac{3}{7} =$

If no order of operation rule applies, calculate from left to right.

2 **Calculate.** 5 points per question

(1) $12 \div (-3) \times (-6) \div (-2)$

$=$

(2) $12 \div [(-3) \times (-6)] \div (-2)$

$=$

(3) $12 \div [(-3) \times (-6) \div (-2)]$

$=$

(4) $-2 \div (-6) \div \frac{1}{2} \div \frac{3}{4}$

$=$

(5) $[-2 \div (-6)] \div \left(\frac{1}{2} \div \frac{3}{4}\right)$

$=$

(6) $-2 \div \left[(-6) \div \frac{1}{2} \div \frac{3}{4}\right]$

$=$

(7) $4\frac{1}{2} \times \left(-1\frac{2}{3}\right) \div 3 \div \frac{3}{4}$

$=$

(8) $4\frac{1}{2} \times \left(-1\frac{2}{3}\right) \div \left(3 \div \frac{3}{4}\right)$

$=$

(9) $4\frac{1}{2} \times \left[\left(-1\frac{2}{3}\right) \div 3\right] \div \frac{3}{4}$

$=$

(10) $(-3)(-4) \div 2 \div 6 \times (-2) \div (-4)$

$=$

(11) $(-3)(-4) \div (2 \div 6) \times (-2) \div (-4)$

$=$

(12) $-3 \times [(-4) \div 2] \div 6 \times (-2) \div (-4)$

$=$

(13) $(-3) \times [(-4) \div 2] \div [6 \times (-2) \div (-4)]$

$=$

(14) $(-3)(-4) \div [2 \div 6 \times (-2)] \div (-4)$

$=$

Positively brilliant work!

25 Negative Numbers with Exponents

Level ★★

Date / / Name

Score /100

■ The Answer Key is on page 189.

1 Determine the sign of the answer. Then calculate.

2 points per question

(1) $(-2)^2 =$

(6) $\left(-\frac{2}{3}\right)^4 =$

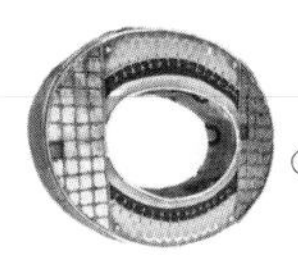

Count the number of negative signs first. Then determine the answer.

(2) $(-2)^3 =$

(7) $\left(-\frac{2}{3}\right)^5 =$

(3) $(-2)^4 =$

(8) $\left(-1\frac{2}{3}\right)^2 =$

(4) $(-2)^5 =$

(9) $-\left(-1\frac{2}{3}\right)^2 =$

(5) $\left(-\frac{2}{3}\right)^3 =$

(10) $\left(-1\frac{2}{3}\right)^3 =$

2 Calculate.

4 points per question

Don't forget!

Note the differences in the examples below.

$4^2=4\times4=16$	$4^3=4\times4\times4=64$
$-4^2=-(4^2)=-(4\times4)=-16$	$-4^3=-(4^3)=-(4\times4\times4)=-64$
$(-4)^2=(-4)\times(-4)=16$	$(-4)^3=(-4)\times(-4)\times(-4)=-64$
$-(-4)^2=-[(-4)\times(-4)]=-16$	$-(-4)^3=-[(-4)\times(-4)\times(-4)]=64$

Pay close attention to where the negative signs are placed.

(1) $2^3=$

(2) $-2^3=-(2^3)=$

(3) $(-2)^3=$

(4) $-(-2)^3=$

(5) $-2^4=$

(6) $(-2)^4=$

(7) $(-4)^2=$

(8) $-4^2=$

(9) $-(-4)^2=$

(10) $-(-4)^3=$

(11) $-\left(\frac{2}{3}\right)^2=$

(12) $\left(-\frac{2}{3}\right)^2=$

(13) $-\left(\frac{2}{3}\right)^3=$

(14) $-\left(-\frac{2}{3}\right)^3=$

(15) $\left(-\frac{2}{3}\right)^4=$

(16) $\left(-1\frac{1}{3}\right)^2=$

(17) $-\left(1\frac{1}{3}\right)^2=$

(18) $\left(-1\frac{1}{3}\right)^3=$

(19) $-\left(1\frac{1}{3}\right)^3=$

(20) $-\left(-1\frac{1}{3}\right)^3=$

Take a bow!

26 Negative Numbers with Exponents

Level ★★

Date / / Name

Score /100

■ The Answer Key is on page 189.

1 Calculate.

5 points per question

(1) $4^3+2^5=64+\square=$

(2) $3^3+4^2=$

(3) $-2^5+(-2)^3=$

(4) $(-3)^2+4^3=$

(5) $-(-3)^2+4^3=$

(6) $7^2-2^6=$

(7) $(-7)^2+(-2)^6=$

(8) $(-6)^2-3^3=$

(9) $-6^2-3^3=$

(10) $-(-6)^2-(-3)^3=$

2 Calculate.

5 points per question

(1) $2^5-3^3-4^2=$

(2) $-2^5-(-3)^3-(-4)^2=$

(3) $-(-2)^5-(-3)^3-4^2=$

(4) $3^2+(-4)^3-(-2)^2=$

(5) $9^2-(-5)^2+(-3)^3=$

(6) $-\left(\frac{1}{3}\right)^2+(-3)^3-\left(\frac{1}{2}\right)^2$

$=$

(7) $\left(-\frac{2}{3}\right)^2+(-2)^4+\left(-\frac{1}{3}\right)^3$

$=$

(8) $-\left(\frac{1}{2}\right)^4-\left(-\frac{1}{3}\right)^2+\left(-\frac{3}{4}\right)^2$

$=$

(9) $-3^3-\left(-1\frac{1}{2}\right)^3-\left(\frac{1}{3}\right)^2$

$=$

(10) $-\left(2\frac{1}{2}\right)^3+\left(-\frac{1}{2}\right)^4-\left(-\frac{1}{3}\right)^2$

$=$

27 Negative Numbers with Exponents

Level 

Date / / Name

Score /100

■ The Answer Key is on page 189.

1 Multiply.

5 points per question

Don't forget!

Determine the sign of the answer before calculating.

$$(-3)^2 \times (-2)^3 = -(3 \times 3 \times 2 \times 2 \times 2) = -72$$

(1) $(-2)^5 \times (-3)^1$

$= 2 \times 2 \times 2 \times 2 \times 2 \times 3$

$=$

(2) $(-3)^3 \times (-2)^2 =$

(3) $(-1)^6 \times (-3)^5 =$

(4) $(-4)^3 \times 2^1 =$

(5) $5^3 \times (-2)^2 =$

(6) $-4^2 \times \left(-\frac{1}{3}\right)^4$

$= \square \left(4 \times \square \times \frac{1}{3} \times \frac{1}{3} \times \frac{1}{3} \times \frac{1}{\square}\right) =$

(7) $3^2 \times \left(-\frac{1}{2}\right)^4 =$

(8) $(-2)^4 \times \left(-\frac{1}{2}\right)^3 =$

(9) $(-6)^2 \times \left(-\frac{2}{3}\right)^3 =$

(10) $-(-2)^3 \times \left(-1\frac{1}{4}\right)^2 =$

Reduce as you calculate!

2 Divide.

5 points per question

Don't forget!

Determine the sign of the answer before calculating.

$$(-4)^3 \div (-5)^2 = -\frac{4 \cdot 4 \cdot 4}{5 \cdot 5} = -\frac{64}{25}$$

$\frac{4 \cdot 4 \cdot 4}{5 \cdot 5}$ means the same as $\frac{4 \times 4 \times 4}{5 \times 5}$

(1) $(-3)^4 \div (-2)^3 =$

(2) $-4^3 \div (-4)^5 =$

(3) $(-3)^4 \div 4^3 =$

(4) $-6^4 \div (-2)^6 =$

(5) $12^2 \div (-6)^5 =$

(6) $6^2 \div \left(-\frac{1}{2}\right)^3 = 6^2 \times (-\square)^3$

$=$

(7) $-4^3 \div \left(-\frac{1}{2}\right)^2 =$

(8) $(-8)^4 \div \left(-\frac{4}{5}\right)^2 =$

(9) $\left(-\frac{2}{3}\right)^3 \div (-6)^3 =$

(10) $-\left(-\frac{3}{4}\right)^4 \div \left(-1\frac{1}{2}\right)^2 =$

28 Operations with Negative Numbers

Level 

Date / / Name

Score /100

■ The Answer Key is on page 189.

1 **Calculate. Write the intermediate steps.** 5 points per question

Don't forget!

When adding or subtracting multiple positive and negative numbers, it is sometimes easier to combine numbers that have the same sign.

Example $6-5-3+4=10-8=2$

(1) $2-6-3+10=12-9=$

(2) $6-8-5+1=$

(3) $-\frac{1}{6}+\frac{1}{4}-\frac{2}{3}=$

(4) $-\frac{1}{2}+\frac{1}{3}-\frac{1}{4}=$

(5) $-9+3\frac{1}{2}+6-2\frac{1}{3}=$

(6) $-\left(-1\frac{2}{5}\right)-\frac{3}{4}+1\frac{1}{2}=$

(7) $-\left(-\frac{1}{2}\right)-\frac{1}{3}-\left(-\frac{1}{4}\right)=$

(8) $-2\frac{3}{4}-\left(-1\frac{5}{6}\right)-3\frac{1}{2}=$

2 Calculate. Write the intermediate steps.

6 points per question

Don't forget!

When adding or subtracting multiple positive or negative fractions, it is sometimes easier to combine fractions that have the same denominator.

Example $-\frac{1}{5}+\frac{1}{3}-\frac{3}{5}=\frac{1}{3}-\frac{4}{5}=-\frac{7}{15}$

(1) $\frac{3}{5}+\frac{1}{6}-\frac{1}{5}=$

(2) $7-1\frac{1}{2}+3\frac{1}{2}=$

(3) $8-\frac{4}{5}-6=$

(4) $1\frac{1}{4}-\frac{1}{3}-\frac{3}{4}=$

(5) $-\frac{2}{5}+\frac{1}{2}+1\frac{4}{5}=$

(6) $-\left(-\frac{3}{8}\right)+\frac{1}{6}-\frac{5}{6}=$

(7) $-\frac{1}{4}-\left(-1\frac{3}{4}\right)+\frac{2}{5}=$

(8) $-\left(-\frac{3}{8}\right)+\left(-\frac{1}{4}\right)-1\frac{7}{8}$

$=$

(9) $8\frac{1}{3}-6-(-3)-1\frac{2}{3}$

$=$

(10) $2\frac{2}{5}-\frac{3}{4}-\left(-6\frac{4}{5}\right)-1\frac{1}{4}$

$=$

You can do it all!

29 Operations with Negative Numbers

Level ☆☆☆ 

Date / / Name

Score 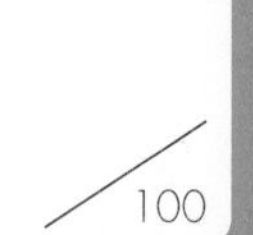 /100

■ The Answer Key is on page 190.

1 Calculate.

5 points per question

Examples

$\frac{2}{3} = 2 \div 3$

$\frac{\frac{1}{2}}{\frac{3}{5}} = \frac{1}{2} \div \frac{3}{5} = \frac{1}{2} \times \frac{5}{3} = \frac{5}{6}$

$\frac{3}{1-\frac{1}{7}} = 3 \div \left(1-\frac{1}{7}\right) = 3 \div \frac{6}{7} = 3 \times \frac{7}{6} = 3\frac{1}{2}$

If there is an operation in the numerator or denominator, place parentheses around it to calculate in the correct order of operations.

(1) $\frac{\frac{1}{3}}{\frac{1}{2}} =$

(2) $\frac{\frac{3}{7}}{\frac{4}{9}} =$

(3) $\frac{\frac{1}{2}}{\frac{3}{4}} =$

(4) $\frac{5}{\frac{1}{2}} =$

(5) $\frac{1\frac{1}{2}}{\frac{1}{3}} =$

(6) $\frac{3}{2-1\frac{3}{4}} = 3 \div \left(2-\square\right) =$

(7) $\frac{3\frac{1}{3}-\frac{1}{6}}{2} =$

(8) $\frac{\frac{1}{2}+\frac{1}{3}}{-\frac{1}{4}+\frac{1}{5}} =$

2 Calculate.

6 points per question

(1) $3 \div 2 - 6 =$

(2) $\dfrac{3}{2-6} =$

(3) $-4 \times 3 - \left(-\dfrac{1}{2}\right) =$

(4) $-(-5) - \dfrac{1}{2} \div (-4) =$

(5) $6 \times \dfrac{1}{2} - (-27) \div (-12)$

$=$

(6) $\dfrac{1}{4} - \left(-\dfrac{1}{2}\right) \div 1\dfrac{1}{2} - \dfrac{1}{6}$

$=$

(7) $\dfrac{\dfrac{1}{4} + \dfrac{1}{3}}{1\dfrac{1}{2} - \dfrac{1}{6}} =$

(8) $-5 \times 3 + 0 \div \left(1\dfrac{1}{2}\right) - \left(-\dfrac{1}{4}\right)$

$=$

(9) $-2\dfrac{1}{2} - \left[\left(-\dfrac{3}{4}\right) \times 2\right] \div \dfrac{3}{5}$

$=$

(10) $-(-2)\left(1\dfrac{1}{4}\right)\left(-1\dfrac{1}{2}\right) - \dfrac{\frac{2}{5}}{4}$

$=$

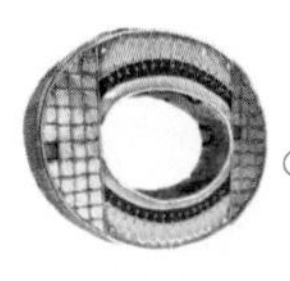

Don't forget to place parentheses around the numerator or denominator if necessary.

Amazing math skills!

30 Operations with Negative Numbers

Level ★★★

Date / / Name

Score /100

■ The Answer Key is on page 190.

1 Calculate.

6 points per question

(1) $(-3)^2(-4)+6\div(-1)=$

(2) $9\div(-6)^2\times(-4)^3-3^0=$

(3) $4^2\times(-5)^2\div(-2)^4=$

(4) $-(-1)^3\times 3^2\times(-2)^3+(-3)^2$

$=$

(5) $\dfrac{6+(-3)^2}{-3-(-2)^2}=$

(6) $\left(-\dfrac{2}{3}\right)^3\div(-4)^2+\left(\dfrac{1}{9}\right)^2(-3)^3$

$=$

(7) $\dfrac{1}{2}-(-1)^3\div\left(\dfrac{1}{3}\right)^3-3$

$=$

(8) $3\dfrac{1}{2}+(-2)^4+(-3)^1\times 0.75$

$=$

(9) $-\left(-\dfrac{3}{5}\right)\left(1\dfrac{5}{6}\right)\left(-\dfrac{4}{7}\right)-2.3$

$=$

(10) $3\dfrac{1}{2}-(-0.8)-5+0.25$

$=$

Convert decimals to fractions if it makes the question easier to calculate.

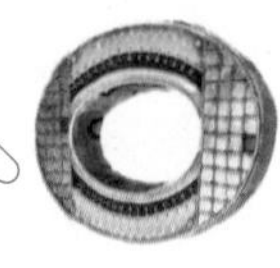

2 Calculate.

5 points per question

(1) $-\left(-2\frac{2}{3}\right)^2+\left(\frac{1}{4}\right)^2-\left(\frac{1}{3}\right)^2$

$=$

(2) $(-3)^4\times\left(-\frac{5}{6}\right)^3=$

(3) $(-1.25)^2\div1\frac{1}{9}\div\left(-1\frac{1}{8}\right)$

$=$

(4) $\dfrac{3\div4}{-\frac{1}{6}+1}=$

(5) $\dfrac{-1\frac{1}{2}+4\frac{2}{3}}{\frac{1}{4}-5}=$

(6) $\dfrac{-6^2\div3^3}{\left(\frac{1}{2}\right)^2\times\left(-1\frac{1}{3}\right)^3}=$

(7) $\left(\dfrac{\frac{2}{3}}{\frac{5}{6}}\right)^2-\dfrac{1-\frac{2}{5}}{6}=$

(8) $-(-0.4)^2+(0.25)(-0.75)\div(0.5)^2$

$=$

You have a powerful mind!

31 Values of Algebraic Expressions

Level 

Date / /

Name

Score /100

■ The Answer Key is on page 190.

1 **Determine the value of each expression when $x=2$.**

3 points per question

Examples

$x+3=2+3=5$ $x-6=2-6=-4$

(1) $x+1=\square+1=$

(2) $x-2=$

(3) $x+10=$

(4) $x-8=$

(5) $4+x=$

(6) $9-x=$

2 **Determine the value of each expression when $x=-3$.**

4 points per question

(1) $x+5=-3+5=$

(2) $x-6=$

(3) $x+1=$

(4) $x-8=$

(5) $4+x=$

(6) $5-x=5-(-3)=$

(7) $-2-x=$

3 Determine the value of each expression when $y=3$.

3 points per question

(1) $2y=2\times\square=$

(2) $5y=$

(3) $3y=$

(4) $4y=$

(5) $-y=$

(6) $\frac{y}{3}=\square\div3=$

(7) $\frac{y}{12}=$

(8) $\frac{6}{y}=$

(9) $\frac{9}{y}=$

(10) $\frac{5}{y}=$

$2y$ means the same as $2\times y$.

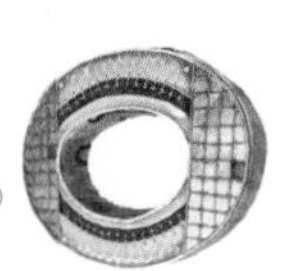

4 Determine the value of each expression when $z=-6$.

3 points per question

(1) $4z=4\times(-6)=$

(2) $7z=$

(3) $6z=$

(4) $10z=$

(5) $\frac{z}{2}=$

(6) $\frac{z}{4}=$

(7) $\frac{12}{z}=$

(8) $\frac{2}{z}=$

Your work is quite valuable!

32 Values of Algebraic Expressions

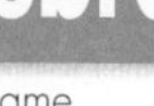

Level ☆☆

Date / / Name

Score /100

■ The Answer Key is on page 190.

1 Determine the value of each expression when $x=6$.

4 points per question

(1) $\frac{x}{2}=$

(4) $\frac{3x}{4}=$

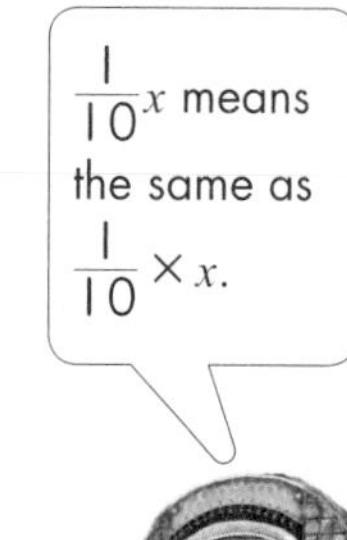

$\frac{1}{10}x$ means
the same as
$\frac{1}{10}\times x$.

(2) $\frac{1}{10}x=$

(5) $-\frac{5}{8}x=$

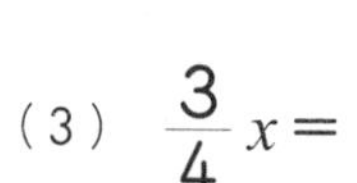

(3) $\frac{3}{4}x=$

(6) $-\frac{11}{6}x=$

2 Determine the value of each expression when $z=8$.

3 points per question

(1) $\frac{z}{2}+3=\frac{\square}{2}+3=$

(3) $\frac{z}{10}-3=$

(2) $-\frac{z}{4}-5=$

(4) $-\frac{5}{6}z+\frac{1}{2}=$

3 Determine the value of each expression when $a=-3$.

4 points per question

(1) $2a+1=$

(2) $-a-2=$

(3) $3a-\frac{1}{3}=$

(4) $7a+\frac{4}{5}=$

(5) $\frac{a}{3}+6=$

(6) $\frac{a}{2}-1=$

(7) $-\frac{2a}{5}-\frac{3}{4}=$

(8) $\frac{5}{4}a-\frac{1}{3}=$

4 Determine the value of each expression when $c=-\frac{1}{2}$.

4 points per question

(1) $6c+4=$

(2) $4c-3=$

(3) $-c+2=$

(4) $2c+1=$

(5) $\frac{c}{3}+4=$

(6) $-\frac{c}{2}-1=$

(7) $\frac{3}{4c}-\frac{3}{4}=$

(8) $-\frac{6}{5}c+2\frac{2}{3}=$

Very good!

33 Values of Algebraic Expressions

Level ★★

Date / /

Name

■ The Answer Key is on page 190.

1 **Determine the value of each expression when $a=4$. Answer with decimals.**

5 points per question

(1) $0.6a=0.6\times\square=$

(2) $-1.4a=$

(3) $0.8a-1.5=$

(4) $-2.7a+0.2=$

(5) $0.03a-0.1=$

(6) $3.5-1.5a=$

(7) $9-0.8a=$

(8) $-12+4.02a=$

(9) $1.01a-5=$

(10) $-3.62a+5.8=$

2 Determine the value of each expression when $b=\frac{1}{2}$.

5 points per question

(1) $-3b=$

(6) $-\frac{3}{4}-\frac{b}{8}=$

(2) $2b-3=$

(7) $3b-\frac{4}{5}=$

(3) $-b+5=$

(8) $-\frac{1}{2}-4b=$

(4) $9b-\frac{1}{2}=$

(9) $-2+3b=$

(5) $-7b+5\frac{1}{4}=$

(10) $\frac{1}{2}-5b=$

34 Values of Algebraic Expressions

Level ★★

Date / /

Name

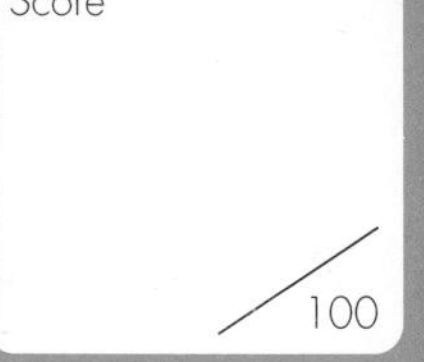

Score /100

■ The Answer Key is on page 191.

1 **Determine the value of each expression when $x = 3$.**

5 points per question

(1) $x^3 = \boxed{3}^{3} =$

(2) $x^4 = \boxed{3}^{4} =$

(3) $-x^2 =$

(4) $x^2 - 3 =$

(5) $x^3 + 8 =$

(6) $10 - x^2 =$

(7) $8 - x^2 =$

(8) $x^3 - 5 =$

(9) $7 - x^2 =$

(10) $-x^4 + 50 =$

2 **Determine the value of each expression when $k=\frac{2}{3}$.** 5 points per question

(1) $k^2=$

(2) $k^3=$

(3) $-3k+1=$

(4) $\frac{9}{2}k-\frac{3}{4}=$

(5) $\frac{5}{6k}+\frac{1}{2}=$

(6) $(2k)^2=\left(2\times\square\right)^2=$

(7) $(-2k)^2=$

(8) $-(-2k)^2=$

(9) $k^2+1=$

(10) $k^2-\frac{1}{3}=$

Awesome!

35 Word Problems with Algebraic Expressions

Level ★★★

Date / /

Name

Score /100

■ The Answer Key is on page 191.

1 **Alex sells 4 pounds of coffee each day. Let x equal the number of days that Alex sells coffee, and answer each word problem.**

10 points per question

(1) Express the number of pounds of coffee that Alex sells in x days.

〈**Ans.**〉

(2) How many pounds of coffee does Alex sell in 6 days?

〈**Ans.**〉

2 **Maria has a party, and she estimates that each guest will eat $\frac{2}{3}$ of a small pizza. Let x equal the number of guests that she invites to the party, and answer each word problem.**

10 points per question

(1) Express the number of pizzas that are eaten by x guests.

〈**Ans.**〉

(2) How many pizzas will be eaten if 5 guests attend and an additional 2 pizzas are eaten by Maria's family? Write the question as an expression first, and then calculate.

〈**Ans.**〉

3 **Bruce puts 1.2 gallons of gas in his car each day. Let x equal the number of days that he puts gas in his car, and answer each word problem. Answer with decimals.**

10 points per question

(1) Express the number of gallons of gas that Bruce puts in his car in x days.

〈Ans.〉

(2) Bruce puts gas in his car each day for 9 days, but he also uses 8.5 gallons of gas. How many gallons of gas remain in his car? Write the question as an expression first, and then calculate.

〈Ans.〉

4 **John builds a square box, therefore the length and width are equal. Let x equal the length and width, and answer each word problem.**

10 points per question

(1) To calculate the area of the box, John multiplies the length by the width.
Express the area of the box.

〈Ans.〉

(2) What is the area of the box when the length and width each equal 4 cm? Answer in cm^2.

〈Ans.〉

(3) John uses the box to build a cube. Therefore, the height is equal to the length and width. To calculate the volume, John multiplies the height by the length by the width. Express the volume of the cube.

〈Ans.〉

(4) What is the volume of the cube when the length, width, and height each equal 4 cm? Answer in cm^3.

〈Ans.〉

Smart thinking!

36 Values of Algebraic Expressions

Level 

Date / /

Name

Score /100

■ The Answer Key is on page 191.

1 Determine the value of each expression when $x=2$.

5 points per question

(1) $x^3+x=\square^3+2=$

(6) $2x^2-\frac{3}{x}=$

(2) $x^2+x=$

(7) $5x^3+\frac{1}{x^4}=$

(3) $x^3-x^2=$

(8) $\frac{1}{x}+\frac{1}{x^2}=$

(4) $3x^4-x=$

(9) $x^2+3x+6=$

(5) $\frac{1}{2}x^3-x^2=$

(10) $20-x^4+3x=$

2 Determine the value of each expression when $x = -3$.

5 points per question

(1) $x^2 - x =$

(6) $\frac{6}{x} - \frac{3}{x^2} =$

(2) $x^3 - x^2 =$

(7) $\frac{1}{x^3 + x} =$

(3) $x^4 - x^3 =$

(8) $\frac{3x^2 - 1}{x^4} =$

(4) $\frac{2x^3}{5} + x^2 =$

(9) $\frac{2 - 4x^2}{x^3 + 3} =$

(5) $\frac{1}{x} - 4 =$

(10) $\frac{x^3 - 5}{x^2 + 1} =$

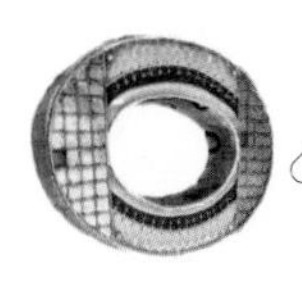

Pay close attention to the negative signs.

37 Values of Algebraic Expressions

Level 

Date / /

Name

Score

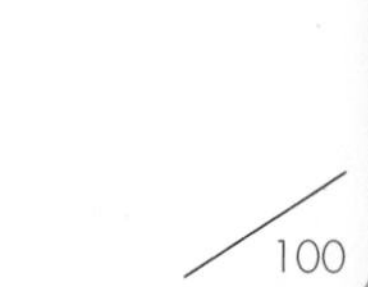

■ The Answer Key is on page 191.

1 **Determine the value of each expression when $d=\frac{1}{2}$.**

5 points per question

(1) $d^2-4d=$

(6) $\frac{1}{d}+3d=$

(2) $d^3+d=$

(7) $\frac{d+3}{d^2-4}=$

(3) $d^2-d^4=$

(8) $\frac{1}{4d^2+d^3}=$

(4) $2d+d^3=$

(9) $5-\frac{1}{d^2}+\frac{d^3}{3}=$

(5) $(3d)^2+d^3=$

(10) $\frac{1}{2d}-\frac{9d}{4}=$

2 Determine the value of each expression when $h=-\frac{3}{2}$.

5 points per question

(1) $\frac{h}{2}+h=$

(6) $h^2+h+\frac{1}{4}=$

(2) $\frac{h}{4}-\frac{h}{2}=$

(7) $\frac{1}{h}-\frac{2}{h^2}=$

(3) $h^2+h=$

(8) $\frac{5}{h}+\frac{1}{h^3}=$

(4) $h-\frac{h^3}{4}=$

(9) $20-\frac{1}{h^2}+\frac{1}{2}h=$

(5) $2h^2+h-5=$

(10) $3h^2-2h+2=$

These are tough. Keep up the good work!

38 Values of Algebraic Expressions

Level 

Date / / Name

Score /100

■ The Answer Key is on page 191.

1 **Determine the value of each expression when $x=2$ and $y=3$.** 5 points per question

Example

$4x-y=4\times2-3=8-3=5$

(1) $x+y=\square+3=$

(2) $x-5y=$

(3) $3x-2y=$

(4) $\frac{1}{2}x+3y=$

(5) $4x-\frac{3}{4}y=$

(6) $\frac{9}{4}x-\frac{5}{2}y=$

(7) $-3y-\frac{3}{7}x=-3\times\square-\frac{3}{7}\times2=$

(8) $y^2-x^2=$

(9) $x^2+y=$

(10) $-x^3-y^2+1=$

2 Determine the value of each expression when $a=-2$ and $b=4$.

5 points per question

(1) $a-b=$

(2) $-a+3b=$

(3) $-5a-ab=$

(4) $3a-\frac{1}{4}b=$

(5) $\frac{5}{8}a+\frac{1}{2}b=$

(6) $-\frac{9}{2}a-\frac{5}{12}b=$

(7) $-\frac{9}{2a}-\frac{5}{12b}=$

(8) $a^2-\frac{a}{b}=$

(9) $a^3b^2=(-2)^3\times4^2=$

(10) $\frac{a^4+3}{(b-6)^2}=$

39 Values of Algebraic Expressions

Level  

Date / /

Name

Score /100

■ The Answer Key is on page 191.

1 **Determine the value of each expression when $x=-3$ and $z=\frac{1}{2}$.** 5 points per question

(1) $x-3z=$

(6) $\frac{xz}{z-1}=$

(2) $x-z=$

(7) $\frac{x-1}{1-z}=$

(3) $z-x=$

(8) $x^2z^2=$

(4) $-(-x)+z=$

(9) $z(x-1)=\frac{1}{2}\times(-3-1)=$

(5) $\frac{x}{z}=$

(10) $\frac{z^3-2}{x+z^2}=$

2 Determine the value of each expression when $f=-\frac{1}{4}$ and $g=\frac{3}{2}$.

5 points per question

(1) $-f+g=$

(2) $-f+(-g)=$

(3) $2f-3g=$

(4) $\frac{f}{g}=$

(5) $\frac{g}{f}=$

(6) $\frac{f+g}{g+f}=$

(7) $f^2-g^2=$

(8) $f+g=$

(9) $f-g=$

(10) $(f+g)\times(f-g)=$

40 Values of Algebraic Expressions

Date / / Name

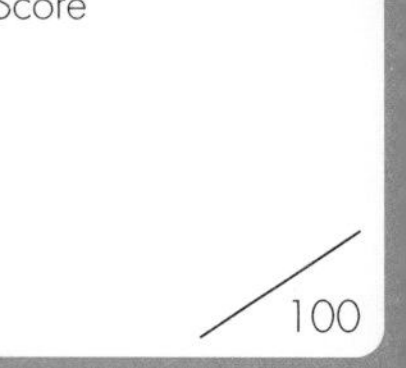

■ The Answer Key is on page 192.

1 **Determine the value of each expression when $x=1$, $y=2$, and $z=3$.** 5 points per question

(1) $x+y-z=1+\square-\square=$

(2) $x-2y+3z=$

(3) $xyz=$

(4) $xy-yz=$

(5) $\dfrac{xy}{z}=$

(6) $\dfrac{x}{y}+\dfrac{y}{z}=$

(7) $\dfrac{x+1}{y-z}=$

(8) $x^2+y^2+z^2=$

(9) $(xy)^2-z^2=$

(10) $(xy+z)(xy-z)=$

2 **Determine the value of each expression when $a=2$, $b=-1$, and $c=-4$.** 5 points per question

(1) $a+b+c=$

(2) $-3a-b+4c=$

(3) $ab-ac=$

(4) $a(b-c)=$

(5) $\frac{c-a}{b}=$

(6) $\frac{1}{b}-\frac{1}{c}=$

(7) $\frac{c-b}{bc}=$

(8) $a^2-b+c^2=$

(9) $(abc)^2=$

(10) $\frac{a^4-b^2}{c^3}=$

You're a math wiz!

Values of Algebraic Expressions

Level ☆☆☆

Date / /

Name

Score /100

■ The Answer Key is on page 192.

1 **Determine the value of each expression when $j=2$, $k=1$, and $l=\frac{1}{2}$.** 5 points per question

(1) $-j+k-l=$

(6) $\dfrac{kl}{j+k}=$

(2) $jk+lj=$

(7) $(jkl)^2=$

(3) $j(k+l)=$

(8) $j^2k^2l^2=$

(4) $\dfrac{\frac{k}{l}}{j}=$

(9) $\dfrac{4j^2}{9k^2}\times l^4=$

(5) $\dfrac{1}{l}+\dfrac{1}{k}+\dfrac{1}{j}=$

(10) $\left(\dfrac{2j}{3k}\times l^2\right)^2=$

2 **Use the given values of a, b, and c to determine the expression $a-(-b-c)$.**

5 points per question

Example

$a=1,\ b=2,\ c=3$

$a-(-b-c)=1-(-2-3)=1-(-5)=6$

(1) $a=3,\ b=5,\ c=-4$

$a-(-b-c)=$

(2) $a=-1,\ b=0,\ c=2$

$a-(-b-c)=$

(3) $a=-2,\ b=0,\ c=2$

$a-(-b-c)=$

(4) $a=1,\ b=\frac{1}{2},\ c=2$

$a-(-b-c)=$

(5) $a=-\frac{1}{2},\ b=\frac{3}{4},\ c=-1$

$a-(-b-c)=$

(6) $a=\frac{3}{4},\ b=-\frac{1}{2},\ c=1$

$a-(-b-c)=$

3 **Use the given values of x, y, and z to determine the expression $\frac{1}{x^2}-\frac{x}{y+z}$.**

5 points per question

(1) $x=1,\ y=2,\ z=3$

$\frac{1}{x^2}-\frac{x}{y+z}=$

(2) $x=-3,\ y=-1,\ z=4$

$\frac{1}{x^2}-\frac{x}{y+z}=$

(3) $x=1,\ y=-2,\ z=-1$

$\frac{1}{x^2}-\frac{x}{y+z}=$

(4) $x=\frac{1}{2},\ y=1,\ z=1$

$\frac{1}{x^2}-\frac{x}{y+z}=$

You get a gold medal in math!

Review

Date / /

Name

Level

Score /100

■ The Answer Key is on page 192.

1 Calculate.

5 points per question

(1) $2^6 \times \frac{1}{2^2} =$

(2) $3^2 \times 3^4 =$

(3) $\left(\frac{2}{3}\right)^1 \times \left(\frac{2}{3}\right)^2 \times \left(\frac{2}{3}\right)^2 =$

(4) $\frac{2^4}{5^2} \times \frac{2^2}{5^1} =$

(5) $6^{10} \div 6^5 \div 6^2 =$

(6) $\left(1\frac{1}{3}\right)^7 \div \left(1\frac{1}{3}\right)^4 =$

(7) $8^4 \div 24^4 =$

(8) $\left(\frac{2}{5}\right)^4 \div 4^4 =$

(9) $\left(4\frac{1}{2}\right)^3 \div \left(\frac{3}{4}\right)^3 =$

(10) $\left(3\frac{1}{3}\right)^2 \times \left(1\frac{4}{5}\right)^2 \div \left(1\frac{1}{2}\right)^2 =$

2 Calculate.

5 points per question

(1) $6+2\times4-3=$

(2) $(6+2)\times(4-3)=$

(3) $6+[2\times(4-3)]=$

(4) $\frac{1}{2}\div\frac{3}{2}+\frac{3}{4}\times6=$

(5) $\frac{1}{2}\div\left(\frac{3}{2}+\frac{3}{4}\right)\times6$

$=$

(6) $\frac{1}{2}\div\left[\frac{3}{2}+\left(\frac{3}{4}\times6\right)\right]$

$=$

(7) $3\div1+2^4\times\frac{1}{2}-2^2$

$=$

(8) $3\div(1+2^4)\times\frac{1}{2}-2^2$

$=$

(9) $3\div\left[\left(1+2^4\times\frac{1}{2}\right)-2^2\right]$

$=$

(10) $(3\div1)+\left[2^4\times\left(2^2-\frac{1}{2}\right)\right]$

$=$

You're almost at the finish line!

Review

Date / /

Name

Level ★★

Score /100

■ The Answer Key is on page 192.

1 **Calculate.**

5 points per question

(1) $6 \div (-2)^3 \times 3^2 - 2^0$

$=$

(2) $(-4)^3 \div (-2)^5 \div 3^2 + (-5)^2$

$=$

(3) $\left(2\frac{2}{3}\right)^2 + \left(-\frac{1}{6}\right) \times \left(\frac{1}{2}\right)$

$=$

(4) $\left(-\frac{2}{3}\right)^4 \div \left(\frac{4}{9}\right)^2 + \left(\frac{1}{6}\right)^2 \times \left(-\frac{9}{10}\right)^1$

$=$

(5) $\left(-1\frac{1}{3}\right)^3 \left(\frac{1}{6}\right)^2 \times (-3)^3 \times \left(-\frac{1}{4}\right)^3$

$=$

(6) $(-4)^3 \times \left(\frac{3}{2}\right)^5 =$

(7) $\dfrac{-2^5 \div 4^3}{\left(\frac{3}{4}\right)^2 \times \left(-\frac{2}{3}\right)^3} =$

(8) $\left(\dfrac{\frac{1}{2}}{\frac{3}{4}}\right)^2 - \left(\frac{1}{3} - 1\frac{1}{3} \times \frac{1}{12}\right)^2$

$=$

2 **Determine the value of each expression when $x=2$, $y=-1$, and $z=\frac{2}{3}$.** 10 points per question

(1) $3x-4z=$

(2) $y(y-z)=$

(3) $(x+5y)^2-z^2=$

(4) $(x+5y-z)(x+5y+z)$

$=$

(5) $z^2-y^2+x^2$

$=$

(6) $(z+x)(z-y)-(y+x)(y-x)$

$=$

Congratulations on completing Kumon's *Pre-Algebra Workbook II*!

Answer Key Grades 6-8 Pre-Algebra Workbook II

1 Greatest Common Factor Review
pp 98, 99

1
(1) $\frac{1}{3}$ (2) $\frac{1}{2}$ (3) $\frac{1}{2}$ (4) $\frac{2}{5}$ (5) $\frac{2}{3}$
(6) $\frac{1}{7}$ (7) $\frac{4}{5}$ (8) $\frac{3}{8}$ (9) $\frac{2}{3}$ (10) $\frac{1}{3}$

2
(1) $\frac{1}{2}$ (2) $\frac{1}{9}$ (3) $\frac{1}{3}$ (4) $\frac{3}{5}$ (5) $\frac{1}{3}$ (6) $\frac{1}{6}$ (7) $\frac{1}{2}$ (8) $\frac{5}{6}$ (9) $\frac{2}{7}$ (10) $\frac{1}{2}$
(11) $\frac{1}{5}$ (12) $\frac{3}{5}$ (13) $\frac{6}{7}$ (14) $\frac{4}{9}$ (15) $\frac{2}{7}$ (16) $\frac{3}{8}$ (17) $\frac{5}{6}$ (18) $\frac{1}{2}$ (19) $\frac{1}{2}$ (20) $\frac{2}{5}$

2 Least Common Multiple Review
pp 100, 101

1
(1) 15 (2) 40 (3) 72 (4) 20 (5) 24
(6) 36 (7) 10 (8) 12 (9) 24 (10) 30

2
(1) 10 (2) 28 (3) 12 (4) 18 (5) 15 (6) 35 (7) 56 (8) 66 (9) 36 (10) 60
(11) 48 (12) 40 (13) 80 (14) 42 (15) 70 (16) 54 (17) 60 (18) 120 (19) 66 (20) 60

3 Addition of Fractions Review
pp 102, 103

1
(1) $\frac{7}{8}$ (2) $\frac{9}{10}$ (3) $\frac{2}{3}$ (4) $\frac{4}{5}$ (5) $\frac{4}{7}$
(6) $\frac{19}{20}$ (7) $\frac{24}{35}$ (8) $1\frac{3}{8}$ (9) $1\frac{7}{15}$ (10) $1\frac{1}{6}$

2
(1) $6\frac{5}{8}$ (2) $2\frac{23}{24}$ (3) $5\frac{3}{8}$ (4) $8\frac{5}{12}$ (5) $8\frac{8}{15}$
(6) $6\frac{2}{3}$ (7) $1\frac{1}{12}$ (8) $1\frac{29}{30}$ (9) $4\frac{11}{40}$ (10) $9\frac{17}{20}$

3
(1) $\frac{2}{3}+\frac{1}{4}=\frac{11}{12}$ **Ans.** $\frac{11}{12}$ of the pizza
(2) $1\frac{4}{5}+2\frac{1}{2}+\frac{1}{6}=4\frac{7}{15}$ **Ans.** $4\frac{7}{15}$ miles

4 Subtraction of Fractions Review
pp 104, 105

1
(1) $\frac{3}{8}$ (2) $\frac{1}{6}$ (3) $\frac{3}{7}$ (4) $\frac{11}{35}$ (5) $\frac{7}{36}$
(6) $\frac{1}{20}$ (7) $\frac{17}{24}$ (8) $\frac{13}{40}$ (9) $\frac{2}{15}$ (10) $\frac{2}{35}$

2
(1) $2\frac{3}{10}$ (2) $3\frac{1}{6}$ (3) $3\frac{1}{20}$ (4) $3\frac{8}{15}$
(5) $1-\frac{4}{5}=\frac{\boxed{5}}{5}-\frac{4}{5}=\frac{1}{5}$
(6) $6-\frac{3}{4}=5\frac{\boxed{4}}{4}-\frac{3}{4}=5\frac{1}{4}$
(7) $4\frac{1}{2}$ (8) $7\frac{2}{5}$ (9) $\frac{5}{18}$ (10) $\frac{1}{20}$ (11) $\frac{1}{6}$ (12) $4\frac{5}{9}$

3
$12\frac{1}{2}-2\frac{2}{3}-5\frac{3}{4}=4\frac{1}{12}$ **Ans.** $4\frac{1}{12}$ pounds

5 Multiplication of Fractions Review
pp 106, 107

1
(1) $\frac{3}{20}$ (2) $\frac{8}{21}$ (3) $\frac{7}{12}$ (4) $\frac{4}{35}$ (5) $\frac{1}{8}$
(6) $\frac{12}{35}$ (7) $\frac{11}{21}$ (8) $\frac{6}{7}$ (9) 18 (10) $5\frac{3}{5}$

2
(1) $\frac{8}{45}$ (2) $\frac{2}{33}$ (3) $\frac{4}{9}$ (4) $1\frac{1}{3}$
(5) $1\frac{1}{2}$ (6) 20 (7) $4\frac{1}{2}$ (8) $5\frac{2}{15}$

3
(1) $3\frac{1}{5}\times1\frac{3}{4}=5\frac{3}{5}$ **Ans.** $5\frac{3}{5}$ pounds
(2) $12\times1\frac{5}{6}=22$ **Ans.** 22 pieces

6 Division of Fractions Review
pp 108, 109

1
(1) $\frac{5}{6}$
(2) $\frac{4}{5}$
(3) $\frac{3}{10}$
(4) $1\frac{1}{35}$
(5) $2\frac{4}{9}$
(6) $\frac{1}{12}$
(7) 14
(8) $\frac{6}{35}$
(9) $1\frac{3}{4}$
(10) $1\frac{2}{3}$

2
(1) $\frac{14}{27}$
(2) $1\frac{7}{9}$
(3) $1\frac{1}{2}$
(4) $\frac{5}{14}$
(5) $\frac{3}{5}$
(6) $\frac{81}{100}$
(7) $\frac{1}{6}$
(8) $\frac{1}{11}$

3
(1) $3\frac{1}{2}\div10\frac{1}{4}=\frac{14}{41}$ **Ans.** $\frac{14}{41}$ kilometers
(2) $6\frac{2}{3}\div1\frac{1}{3}\div12=\frac{5}{12}$ **Ans.** $\frac{5}{12}$ ounces

7 Decimals and Fractions
pp 110, 111

1
(1) $\frac{9}{14}$
(2) $1\frac{13}{15}$
(3) 8
(4) $7\frac{7}{24}$
(5) $6\frac{37}{40}$
(6) $\frac{7}{20}$
(7) $4\frac{7}{15}$
(8) $2\frac{3}{8}$
(9) $3\frac{3}{4}$
(10) $\frac{1}{40}$

2
(1) $\frac{1}{5}$
(2) $\frac{2}{3}$
(3) 10
(4) $7\frac{1}{3}$
(5) $7\frac{1}{2}$
(6) $1\frac{2}{3}$
(7) $\frac{5}{9}$
(8) $1\frac{5}{17}$
(9) $1\frac{1}{2}$
(10) $2\frac{2}{3}$

8 Exponent Review
pp 112, 113

1
(1) 8
(2) 9
(3) 2
(4) 27
(5) 32
(6) 1
(7) 1
(8) 216
(9) 343
(10) 128

2
(1) $\frac{1}{8}$
(2) $\frac{1}{16}$
(3) $\frac{16}{81}$
(4) $7\frac{1}{9}$
(5) $2\frac{10}{27}$
(6) $\frac{2}{27}$
(7) $16\frac{1}{5}$
(8) $\frac{8}{81}$
(9) $\frac{27}{49}$
(10) $\frac{64}{729}$

9 Exponent Review
pp 114, 115

1
(1) 243
(2) 64
(3) 16
(4) 256
(5) 625
(6) $\frac{1}{32}$
(7) $\frac{1}{64}$
(8) $\frac{16}{81}$
(9) $\frac{81}{256}$
(10) $\frac{32}{3125}$

2
(1) $2^4\times\left(\frac{1}{2}\right)^2=2\times2\times2\times2\times\frac{1}{2}\times\frac{1}{2}=4$
(2) 25
(3) 64
(4) $\frac{1}{9}$
(5) 50
(6) $20\frac{5}{6}$
(7) $\frac{1}{72}$
(8) $\frac{4}{375}$
(9) $\frac{1}{48}$
(10) $6\frac{2}{9}$

3 $2\frac{1}{2}\times2\frac{1}{2}\times6=37\frac{1}{2}$ **Ans.** $37\frac{1}{2}$ m^3

10 Exponents Multiplication
pp 116, 117

1
(1) $3^3\times3^2=3^5=243$
(2) $4^1\times4^2=4^3=64$
(3) $3^0\times3^4=3^4=81$
(4) $2^3\times2^4=2^7=128$
(5) $5^2\times5^2=5^4=625$
(6) $\frac{1}{32}$
(7) $\frac{1}{256}$
(8) $\frac{8}{27}$
(9) $3\frac{1}{16}$
(10) $4\frac{21}{25}$

2
(1) $\frac{2^3}{3}\times\frac{2^1}{7}=\frac{2^{\boxed{4}}}{21}=\frac{16}{21}$
(2) $\frac{64}{77}$
(3) $\frac{8}{243}$
(4) $\frac{15}{32}$
(5) 6
(6) $\frac{64}{243}$
(7) $\frac{64}{125}$
(8) $1\frac{1}{8}$
(9) $1\frac{15}{49}$
(10) $\frac{81}{256}$
(11) $\frac{27}{160}$
(12) $\frac{32}{125}$

11 Exponents Division
pp 118, 119

1
(1) $4^3\div4^1=4^2=16$
(2) $3^5\div3^3=3^2=9$
(3) $5^4\div5^1=5^3=125$
(4) $2^6\div2^2=2^4=16$
(5) $7^8\div7^6=7^2=49$
(6) $\frac{1}{81}$
(7) $\frac{1}{16}$
(8) $\frac{3}{4}$
(9) $\frac{8}{125}$
(10) $7\frac{9}{16}$

2 (1) 32

(2) 81

(3) 16

(4) $\frac{1}{8}$

(5) $\frac{2}{3}$

(6) $\frac{8}{125}$

(7) $13\frac{4}{9}$

(8) $4\frac{1}{2}$

(9) 1

(10) $4\frac{21}{25}$

12 Exponents Division

pp 120, 121

1 (1) $12^2 \div 4^2 = \left(\frac{12}{4}\right)^2 = 9$

(2) $6^3 \div 3^3 = \left(\frac{6}{3}\right)^3 = 8$

(3) $15^5 \div 5^5 = \left(\frac{15}{5}\right)^5 = 243$

(4) $100^4 \div 10^4 = \left(\frac{100}{10}\right)^4 = 10{,}000$

(5) $56^2 \div 7^2 = \left(\frac{56}{7}\right)^2 = 64$

(6) $\frac{1}{64}$

(7) $\frac{1}{81}$

(8) $\frac{8}{125}$

(9) $3\frac{3}{8}$

(10) $1\frac{61}{64}$

2 (1) $\left(\frac{1}{3}\right)^3 \div (2)^3 = \left(\frac{1}{3} \div 2\right)^3 = \left(\frac{1}{3} \times \frac{1}{2}\right)^3 = \frac{1}{216}$

(2) $\frac{49}{100}$

(3) $\frac{9}{256}$

(4) $12\frac{1}{4}$

(5) $2\frac{10}{27}$

(6) $\frac{27}{64}$

(7) $\frac{25}{36}$

(8) $5\frac{19}{25}$

(9) $14\frac{1}{16}$

(10) $\frac{1}{1000}$

13 Exponents Review

pp 122, 123

1 (1) 9

(2) $\frac{27}{32}$

(3) $\frac{1}{6}$

(4) $\frac{25}{144}$

(5) 225

(6) 128

(7) $\frac{1}{243}$

(8) 64

(9) $1\frac{3}{125}$

(10) $\frac{64}{405}$

2 (1) 9

(2) $\frac{27}{125}$

(3) $15\frac{5}{8}$

(4) 27

(5) $5\frac{4}{9}$

(6) 81

(7) $\frac{8}{125}$

(8) $4\frac{17}{27}$

(9) $\frac{1}{729}$

(10) $1\frac{7}{9}$

(11) $\frac{25}{49}$

(12) 144

14 Order of Operations Review

pp 124, 125

1 (1) 12

(2) 9

(3) 13

(4) $1\frac{1}{2}$

(5) $\frac{5}{12}$

(6) 12

(7) 6

(8) $4\frac{1}{2}$

(9) $\frac{1}{6}$

(10) $\frac{5}{9}$

2 (1) $3+2\times4=3+\boxed{8}=11$

(2) $8\div4+5=\boxed{2}+5=7$

(3) 8

(4) 3

(5) $7\frac{1}{2}$

(6) 15

(7) 0

(8) $4\frac{3}{4}$

(9) $\frac{11}{15}$

(10) 11

(11) $13\frac{1}{2}$

(12) $1\frac{9}{10}$

3 $5\times4+12\times2=44$

Ans. 44 animals

15 Order of Operations

pp 126, 127

1 (1) $9-3+5=11$

(2) $9-(3+5)=9-\boxed{8}=1$

(3) 3

(4) 3

(5) 3

(6) 2

(7) 5

(8) 9

(9) 5

(10) 4

2 (1) $2\frac{1}{2}$

(2) $1\frac{1}{2}$

(3) $\frac{5}{6}$

(4) $19\frac{3}{4}$

(5) $\frac{1}{5}$

(6) $1\frac{1}{2}$

(7) $1\frac{11}{16}$

(8) $3\frac{1}{4}$

(9) $\frac{5}{8}$

(10) $\frac{5}{32}$

(11) 9

(12) $8\frac{1}{4}$

3 $(5\times2+2\times4)\times15=270$

or $[(5\times15)\times2]+[(2\times15)\times4]=270$

Ans. $270

16 Order of Operations pp 128, 129

1
(1) $5+2^3\times4=5+\boxed{8}\times4=37$
(2) $(5+2^3)\times4=(5+\boxed{8})\times4=\boxed{13}\times4=52$
(3) $4\frac{1}{8}$
(4) $3\frac{1}{4}$
(5) $\frac{4}{9}$
(6) $7\frac{1}{5}$
(7) $5\frac{4}{5}$
(8) 4
(9) $\frac{4}{7}$
(10) $2\frac{1}{9}$

2
(1) $19-2^{3+1}=19-2^{\boxed{4}}=3$
(2) $5\frac{1}{16}$
(3) $4\frac{1}{8}$
(4) 216
(5) $14\frac{13}{20}$
(6) $3\frac{1}{2}$
(7) $\frac{2}{3}$
(8) $\frac{25}{72}$

17 Order of Operations Review pp 130, 131

1
(1) 30
(2) 40
(3) $\frac{5}{8}$
(4) $\frac{11}{20}$
(5) $7\frac{3}{4}$
(6) $\frac{9}{20}$
(7) $\frac{7}{20}$
(8) $6\frac{1}{5}$
(9) $8\frac{1}{5}$
(10) 1

2
(1) $1\frac{1}{72}$
(2) $\frac{37}{48}$
(3) $\frac{3}{8}$
(4) $\frac{67}{96}$
(5) $\frac{19}{24}$
(6) 17
(7) 53
(8) 57
(9) 147
(10) 675

18 Negative Numbers pp 132, 133

1
(1) 2
(2) 1
(3) 0
(4) -1
(5) -2
(6) 2
(7) 0
(8) -1
(9) -3
(10) -4

2
(1) 1
(2) 0
(3) -1
(4) -2
(5) 2
(6) 1
(7) 0
(8) -1
(9) 3
(10) 1
(11) 0
(12) -1
(13) -2
(14) -5
(15) -10
(16) -25

19 Addition with Negative Numbers pp 134, 135

1
(1) 3
(2) 2
(3) 1
(4) 0
(5) -1
(6) 10
(7) 4
(8) 1
(9) 0
(10) -5

2
(1) 9
(2) 4
(3) 6
(4) -1
(5) 1
(6) 7
(7) 2
(8) -4
(9) 0
(10) 1
(11) 2
(12) 3
(13) -4
(14) -1
(15) 4
(16) -5
(17) 0
(18) 12
(19) 80
(20) 200

20 Subtraction with Negative Numbers pp 136, 137

1
(1) 1
(2) 0
(3) -1
(4) -2
(5) -3
(6) -1
(7) -7
(8) 2
(9) -10
(10) -3
(11) -5
(12) -2
(13) -6
(14) -13
(15) -13
(16) -14
(17) -9
(18) -25
(19) -30
(20) -25

2
(1) $\frac{1}{4}$
(2) 0
(3) $0-\frac{1}{4}=-\frac{\boxed{1}}{4}$
(4) $-\frac{1}{4}-\frac{1}{4}=-\frac{\boxed{2}}{4}=-\frac{1}{2}$
(5) $-\frac{3}{4}$
(6) $-\frac{1}{5}$
(7) $-1\frac{1}{5}$
(8) $-2\frac{4}{5}$
(9) $\frac{1}{4}-\frac{1}{2}=\frac{1}{4}-\frac{\boxed{2}}{4}=-\frac{1}{4}$
(10) $-\frac{5}{12}$
(11) $-3\frac{2}{3}$
(12) $-\frac{1}{6}$
(13) $\frac{3}{4}$
(14) $\frac{1}{2}-1\frac{1}{4}=\frac{\boxed{2}}{4}-1\frac{1}{4}$
$=\frac{2}{4}-\frac{\boxed{5}}{4}=-\frac{3}{4}$
(15) $-4\frac{1}{2}$

21 Subtraction with Negative Numbers pp 138, 139

1
(1) $4+(-1)=4\boxed{-}1=3$
(2) $-2+(-4)=-2\boxed{-}4=-6$
(3) -5
(4) $\frac{1}{6}$
(5) $-3\frac{11}{12}$
(6) $2-(-6)=2\boxed{+}6=8$
(7) $-3-(-2)=-3\boxed{+}2=-1$
(8) 4
(9) 1
(10) $1\frac{3}{5}$

2
(1) 7 (11) $\frac{1}{3}$
(2) 3 (12) 1
(3) -3 (13) $-\frac{1}{3}$
(4) -7 (14) -1
(5) 7 (15) $-\frac{1}{3}$
(6) 3 (16) 1
(7) -3 (17) -1
(8) -7 (18) $\frac{1}{3}$
(9) 9 (19) $1\frac{5}{6}$
(10) -9 (20) $-\frac{5}{6}$

22 Multiplication with Negative Numbers pp 140, 141

1
(1) $\boxed{2}$ negative signs
(positive) / negative answer
(2) $\boxed{2}$ negative signs
(positive) / negative answer
(3) $\boxed{3}$ negative signs
positive / (negative) answer
(4) $\boxed{4}$ negative signs
(positive) / negative answer

2
(1) -6 (4) 24
(2) 15 (5) 0
(3) -30 (6) -1

3
(1) -1 (9) $1\frac{1}{12}$
(2) 1 (10) $5\frac{1}{4}$
(3) -1 (11) -30
(4) -12 (12) -24
(5) 36 (13) 16
(6) -6 (14) $-1\frac{1}{2}$
(7) $-2\frac{2}{5}$ (15) $\frac{14}{45}$
(8) $4\frac{4}{5}$ (16) $-\frac{3}{5}$

23 Division with Negative Numbers pp 142, 143

1
(1) -5 (6) -2
(2) 2 (7) -1
(3) 0 (8) 3
(4) -2 (9) -2
(5) 4 (10) -5

2
(1) -16 (11) $-\frac{2}{3}$
(2) -12 (12) $2\frac{3}{4}$
(3) $\frac{1}{3}$ (13) -8
(4) $-2\frac{1}{2}$ (14) 3
(5) $-\frac{10}{27}$ (15) $-\frac{8}{15}$
(6) 10 (16) -6
(7) $-\frac{1}{4}$ (17) $\frac{1}{40}$
(8) -15 (18) $\frac{5}{8}$
(9) $1\frac{7}{8}$ (19) $-1\frac{7}{9}$
(10) $-\frac{1}{2}$ (20) 10

24 Multiplication & Division with Negative Numbers pp 144, 145

1
(1) -1 (6) $-1\frac{1}{10}$
(2) -36 (7) $-\frac{1}{9}$
(3) 0 (8) -4
(4) $-\frac{3}{5}$ (9) 5
(5) $\frac{15}{16}$ (10) $-5\frac{1}{4}$

2
(1) -12 (8) $-1\frac{7}{8}$
(2) $-\frac{1}{3}$ (9) $-3\frac{1}{3}$
(3) $-1\frac{1}{3}$ (10) $\frac{1}{2}$
(4) $\frac{8}{9}$ (11) 18
(5) $\frac{1}{2}$ (12) $\frac{1}{2}$
(6) $\frac{1}{8}$ (13) 2
(7) $-3\frac{1}{3}$ (14) $4\frac{1}{2}$

25 Negative Numbers with Exponents pp 146, 147

1
(1) 4
(2) -8
(3) 16
(4) -32
(5) $-\frac{8}{27}$
(6) $\frac{16}{81}$
(7) $-\frac{32}{243}$
(8) $2\frac{7}{9}$
(9) $-2\frac{7}{9}$
(10) $-4\frac{17}{27}$

2
(1) 8
(2) $-2^3=-(2^3)=-8$
(3) -8
(4) 8
(5) -16
(6) 16
(7) 16
(8) -16
(9) -16
(10) 64
(11) $-\frac{4}{9}$
(12) $\frac{4}{9}$
(13) $-\frac{8}{27}$
(14) $\frac{8}{27}$
(15) $\frac{16}{81}$
(16) $1\frac{7}{9}$
(17) $-1\frac{7}{9}$
(18) $-2\frac{10}{27}$
(19) $-2\frac{10}{27}$
(20) $2\frac{10}{27}$

26 Negative Numbers with Exponents pp 148, 149

1
(1) $4^3+2^5=64+\boxed{32}=96$
(2) 43
(3) -40
(4) 73
(5) 55
(6) -15
(7) 113
(8) 9
(9) -63
(10) -9

2
(1) -11
(2) -21
(3) 43
(4) -59
(5) 29
(6) $-27\frac{13}{36}$
(7) $16\frac{11}{27}$
(8) $\frac{7}{18}$
(9) $-23\frac{53}{72}$
(10) $-15\frac{97}{144}$

27 Negative Numbers with Exponents pp 150, 151

1
(1) $(-2)^5\times(-3)^1=2\times2\times2\times2\times2\times3=96$
(2) -108
(3) -243
(4) -128
(5) 500
(6) $-4^2\times\left(-\frac{1}{3}\right)^4=\boxed{-}\left(4\times\boxed{4}\times\frac{1}{3}\times\frac{1}{3}\times\frac{1}{3}\times\frac{1}{\boxed{3}}\right)=-\frac{16}{81}$
(7) $\frac{9}{16}$
(8) -2
(9) $-10\frac{2}{3}$
(10) $12\frac{1}{2}$

2
(1) $-10\frac{1}{8}$
(2) $\frac{1}{16}$
(3) $1\frac{17}{64}$
(4) $-20\frac{1}{4}$
(5) $-\frac{1}{54}$
(6) $6^2\div\left(-\frac{1}{2}\right)^3=6^2\times(-\boxed{2})^3=-288$
(7) -256
(8) $6{,}400$
(9) $\frac{1}{729}$
(10) $-\frac{9}{64}$

28 Operations with Negative Numbers pp 152, 153

1
(1) $2-6-3+10=12-9=3$
(2) -6
(3) $-\frac{7}{12}$
(4) $-\frac{5}{12}$
(5) $-1\frac{5}{6}$
(6) $2\frac{3}{20}$
(7) $\frac{5}{12}$
(8) $-4\frac{5}{12}$

2
(1) $\frac{17}{30}$
(2) 9
(3) $1\frac{1}{5}$
(4) $\frac{1}{6}$
(5) $1\frac{9}{10}$
(6) $-\frac{7}{24}$
(7) $1\frac{9}{10}$
(8) $-1\frac{3}{4}$
(9) $3\frac{2}{3}$
(10) $7\frac{1}{5}$

29 Operations with Negative Numbers pp 154, 155

1 (1) $\frac{2}{3}$ (2) $\frac{27}{28}$ (3) $\frac{2}{3}$ (4) 10
(5) $4\frac{1}{2}$
(6) $\frac{3}{2-1\frac{3}{4}}=3\div\left(2-\boxed{\frac{7}{4}}\right)=12$
(7) $1\frac{7}{12}$
(8) $-16\frac{2}{3}$

2 (1) $-4\frac{1}{2}$ (2) $-\frac{3}{4}$ (3) $-11\frac{1}{2}$ (4) $5\frac{1}{8}$ (5) $\frac{3}{4}$
(6) $\frac{5}{12}$ (7) $\frac{7}{16}$ (8) $-14\frac{3}{4}$ (9) 0 (10) $-3\frac{17}{20}$

30 Operations with Negative Numbers pp 156, 157

1 (1) -42 (2) -17 (3) 25 (4) -63 (5) $-2\frac{1}{7}$
(6) $-\frac{19}{54}$ (7) $24\frac{1}{2}$ (8) $17\frac{1}{4}$ (9) $-2\frac{13}{14}$ (10) $-\frac{9}{20}$

2 (1) $-7\frac{23}{144}$ (2) $-46\frac{7}{8}$ (3) $-1\frac{1}{4}$ (4) $\frac{9}{10}$
(5) $-\frac{2}{3}$ (6) $2\frac{1}{4}$ (7) $\frac{27}{50}$ (8) $-\frac{91}{100}$

31 Values of Algebraic Expressions pp 158, 159

1 (1) $x+1=\boxed{2}+1=3$ (2) 0 (3) 12
(4) -6 (5) 6 (6) 7

2 (1) 2 (2) -9 (3) -2 (4) -11
(5) 1 (6) $5-x=5-(-3)=8$ (7) 1

3 (1) $2y=2\times\boxed{3}=6$ (2) 15 (3) 9 (4) 12 (5) -3
(6) $y\div3=\boxed{3}\div3=1$ (7) $\frac{1}{4}$ (8) 2 (9) 3 (10) $1\frac{2}{3}$

4 (1) -24 (2) -42 (3) -36 (4) -60
(5) -3 (6) $-1\frac{1}{2}$ (7) -2 (8) $-\frac{1}{3}$

32 Values of Algebraic Expressions pp 160, 161

1 (1) $\frac{x}{2}=\frac{\boxed{6}}{2}=3$ (2) $\frac{3}{5}$ (3) $4\frac{1}{2}$
(4) $4\frac{1}{2}$ (5) $-3\frac{3}{4}$ (6) -11

2 (1) $\frac{z}{2}+3=\frac{\boxed{8}}{2}+3=7$ (2) -7
(3) $-2\frac{1}{5}$ (4) $-6\frac{1}{6}$

3 (1) -5 (2) 1 (3) $-9\frac{1}{3}$ (4) $-20\frac{1}{5}$
(5) 5 (6) $-2\frac{1}{2}$ (7) $\frac{9}{20}$ (8) $-4\frac{1}{12}$

4 (1) 1 (2) -5 (3) $2\frac{1}{2}$ (4) 0
(5) $3\frac{5}{6}$ (6) $-\frac{3}{4}$ (7) $-2\frac{1}{4}$ (8) $3\frac{4}{15}$

33 Values of Algebraic Expressions pp 162, 163

1 (1) $0.6a=0.6\times\boxed{4}=2.4$ (2) -5.6 (3) 1.7 (4) -10.6 (5) 0.02
(6) -2.5 (7) 5.8 (8) 4.08 (9) -0.96 (10) -8.68

2 (1) $-1\frac{1}{2}$ (2) -2 (3) $4\frac{1}{2}$ (4) 4 (5) $1\frac{3}{4}$
(6) $-\frac{13}{16}$ (7) $\frac{7}{10}$ (8) $-2\frac{1}{2}$ (9) $-\frac{1}{2}$ (10) -2

34 Values of Algebraic Expressions
pp 164, 165

1
(1) $x^3 = \boxed{3}^3 = 27$
(2) $x^4 = \boxed{3}^4 = 81$
(3) -9
(4) 6
(5) 35
(6) 1
(7) -1
(8) 22
(9) -2
(10) -31

2
(1) $\frac{4}{9}$
(2) $\frac{8}{27}$
(3) -1
(4) $2\frac{1}{4}$
(5) $1\frac{3}{4}$
(6) $(2k)^2 = \left(2 \times \boxed{\frac{2}{3}}\right)^2 = 1\frac{7}{9}$
(7) $1\frac{7}{9}$
(8) $-1\frac{7}{9}$
(9) $1\frac{4}{9}$
(10) $\frac{1}{9}$

35 Word Problems with Algebraic Expressions
pp 166, 167

1
(1) $4 \times x = 4x$ **Ans.** $4x$
(2) $4 \times 6 = 24$ **Ans.** 24 pounds

2
(1) $\frac{2}{3} \times x = \frac{2}{3}x$ **Ans.** $\frac{2}{3}x$
(2) $\frac{2}{3} \times 5 + 2 = 5\frac{1}{3}$ **Ans.** $5\frac{1}{3}$ pizzas

3
(1) $1.2 \times x = 1.2x$ **Ans.** $1.2x$
(2) $1.2 \times 9 - 8.5 = 2.3$ **Ans.** 2.3 gallons

4
(1) $x \times x = x^2$ **Ans.** x^2
(2) $4^2 = 16$ **Ans.** 16 cm^2
(3) $x \times x \times x = x^3$ **Ans.** x^3
(4) $4^3 = 64$ **Ans.** 64 cm^3

36 Values of Algebraic Expressions
pp 168, 169

1
(1) $x^3 + x = \boxed{2}^3 + 2 = 10$
(2) 6
(3) 4
(4) 46
(5) 0
(6) $6\frac{1}{2}$
(7) $40\frac{1}{16}$
(8) $\frac{3}{4}$
(9) 16
(10) 10

2
(1) 12
(2) -36
(3) 108
(4) $-1\frac{4}{5}$
(5) $-4\frac{1}{3}$
(6) $-2\frac{1}{3}$
(7) $-\frac{1}{30}$
(8) $\frac{26}{81}$
(9) $1\frac{5}{12}$
(10) $-3\frac{1}{5}$

37 Values of Algebraic Expressions
pp 170, 171

1
(1) $-1\frac{3}{4}$
(2) $\frac{5}{8}$
(3) $\frac{3}{16}$
(4) $1\frac{1}{8}$
(5) $2\frac{3}{8}$
(6) $3\frac{1}{2}$
(7) $-\frac{14}{15}$
(8) $\frac{8}{9}$
(9) $1\frac{1}{24}$
(10) $-\frac{1}{8}$

2
(1) $-2\frac{1}{4}$
(2) $\frac{3}{8}$
(3) $\frac{3}{4}$
(4) $-\frac{21}{32}$
(5) -2
(6) 1
(7) $-1\frac{5}{9}$
(8) $-3\frac{17}{27}$
(9) $18\frac{29}{36}$
(10) $11\frac{3}{4}$

38 Values of Algebraic Expressions
pp 172, 173

1
(1) $x + y = \boxed{2} + 3 = 5$
(2) -13
(3) 0
(4) 10
(5) $5\frac{3}{4}$
(6) -3
(7) $-3y - \frac{3}{7}x = -3 \times \boxed{3} - \frac{3}{7} \times 2 = -9\frac{6}{7}$
(8) 5
(9) 7
(10) -16

2
(1) -6
(2) 14
(3) 18
(4) -7
(5) $\frac{3}{4}$
(6) $7\frac{1}{3}$
(7) $2\frac{7}{48}$
(8) $4\frac{1}{2}$
(9) -128
(10) $4\frac{3}{4}$

39 Values of Algebraic Expressions
pp 174, 175

1
(1) $-4\frac{1}{2}$
(2) $-3\frac{1}{2}$
(3) $3\frac{1}{2}$
(4) $-2\frac{1}{2}$
(5) -6
(6) 3
(7) -8
(8) $2\frac{1}{4}$
(9) -2
(10) $\frac{15}{22}$

2
(1) $1\frac{3}{4}$
(2) $-1\frac{1}{4}$
(3) -5
(4) $-\frac{1}{6}$
(5) -6
(6) 1
(7) $-2\frac{3}{16}$
(8) $1\frac{1}{4}$
(9) $-1\frac{3}{4}$
(10) $-2\frac{3}{16}$

40 Values of Algebraic Expressions
pp 176, 177

1
(1) $x+y-z=1+\boxed{2}-\boxed{3}=0$
(2) 6
(3) 6
(4) -4
(5) $\frac{2}{3}$
(6) $1\frac{1}{6}$
(7) -2
(8) 14
(9) -5
(10) -5

2
(1) -3
(2) -21
(3) 6
(4) 6
(5) 6
(6) $-\frac{3}{4}$
(7) $-\frac{3}{4}$
(8) 21
(9) 64
(10) $-\frac{15}{64}$

41 Values of Algebraic Expressions
pp 178, 179

1
(1) $-1\frac{1}{2}$
(2) 3
(3) 3
(4) 1
(5) $3\frac{1}{2}$
(6) $\frac{1}{6}$
(7) 1
(8) 1
(9) $\frac{1}{9}$
(10) $\frac{1}{9}$

2
(1) 4
(2) 1
(3) 0
(4) $3\frac{1}{2}$
(5) $-\frac{3}{4}$
(6) $1\frac{1}{4}$

3
(1) $\frac{4}{5}$
(2) $1\frac{1}{9}$
(3) $1\frac{1}{3}$
(4) $3\frac{3}{4}$

42 Review
pp 180, 181

1
(1) 16
(2) 729
(3) $\frac{32}{243}$
(4) $\frac{64}{125}$
(5) 216
(6) $2\frac{10}{27}$
(7) $\frac{1}{81}$
(8) $\frac{1}{10000}$
(9) 216
(10) 16

2
(1) 11
(2) 8
(3) 8
(4) $4\frac{5}{6}$
(5) $1\frac{1}{3}$
(6) $\frac{1}{12}$
(7) 7
(8) $-3\frac{31}{34}$
(9) $\frac{3}{5}$
(10) 59

43 Review
pp 182, 183

1
(1) $-7\frac{3}{4}$
(2) $25\frac{2}{9}$
(3) $7\frac{1}{36}$
(4) $\frac{39}{40}$
(5) $-\frac{1}{36}$
(6) -486
(7) 3
(8) $\frac{32}{81}$

2
(1) $3\frac{1}{3}$
(2) $1\frac{2}{3}$
(3) $8\frac{5}{9}$
(4) $8\frac{5}{9}$
(5) $3\frac{4}{9}$
(6) $7\frac{4}{9}$

